EXISTENTIAL POLITICS

Existential Politics

WHY GLOBAL CLIMATE INSTITUTIONS
ARE FAILING AND HOW TO FIX THEM

JESSICA F. GREEN

PRINCETON UNIVERSITY PRESS

PRINCETON & OXFORD

Published by Princeton University Press
41 William Street, Princeton, New Jersey 08540
99 Banbury Road, Oxford OX2 6JX

press.princeton.edu

GPSR Authorized Representative: Easy Access System Europe - Mustamäe tee 50, 10621 Tallinn, Estonia, gpsr.requests@easproject.com

All Rights Reserved

ISBN 9780691245232
ISBN (pbk.) 9780691245249
ISBN (e-book) 9780691245256

Library of Congress Control Number: 2025930776

British Library Cataloging-in-Publication Data is available

Editorial: Bridget Flannery-McCoy, Dave McBride, and Alena Chekanov
Production Editorial: Theresa Liu
Jacket/Cover Design: Karl Spurzem
Production: Lauren Reese
Publicity: William Pagdatoon
Copyeditor: Cynthia Buck

THIS book has been composed in Arno

10 9 8 7 6 5 4 3 2 1

For Hannah and Milo

CONTENTS

ACKNOWLEDGMENTS

THIS BOOK began as a paper called "Asset Revaluation and the Existential Politics of Climate Change," coauthored with longtime friends and colleagues, Jeff Colgan and Tom Hale. The paper argues that climate change is not a collective action problem, as economists and political scientists have long insisted. Instead, it is best understood as a political-economic contest between fossil asset owners, who will lose from decarbonization, and vulnerable asset owners, who will lose from the effects of climate change.

After the paper was published, I realized that our thinking was incomplete. The paper outlined a contest between losers. But without winners, prospects for ambitious climate policy are narrower. After I presented this work to a group of climate scholars, Kathryn Sikkink asked me, "What is the model for change?" I didn't have a good answer, which meant, of course, that she was asking the right question. Where was the pathway to decarbonization in our theory of asset revaluation? We were missing the winners—those with material incentives to advocate for more ambitious climate policy. Winners will be the key to accelerating the energy transition.

I started this book to think more deeply about how to cultivate winners as a political interest group. The book builds on the article on existential politics as well as two decades of research and thinking on climate politics. It also incorporates research on net zero emissions with Tom Hale, Raul Salas Reyes and Aldrick Arceo, as well as work on oil companies' climate strategies with Jennifer Hadden, Tom Hale, and Paasha Mahdavi.

Writing this book has shown me, yet again, how lucky I am to be surrounded by amazing friends and colleagues. The book has truly been a team effort—of colleagues, friends, and family. I wrote it during a very strange time in the world, and in my own life. I was on my first research leave in a decade when COVID happened. Leave quickly turned into homeschooling my two young children. Like many others, I faced significant personal challenges during and after the pandemic. This project helped give me purpose and focus during some very difficult times. As my sister says, "Writing is joy." I was fortunate enough to experience this joy firsthand. This book is, in many ways, how I found joy and purpose in a time of grief and fear.

My excellent colleagues also helped tremendously. At the University of Toronto, my two chairs, Ryan Balot and Steve Easterbrook, were supportive of my work and kind enough to ensure that I had leave and accommodation at key moments. As interim chair, Lou Pauly was just as understanding. The crew at the Environmental Governance Lab—especially Steven Bernstein, Matt Hoffmann, and Kate Neville—provided useful perspective on bits and pieces of the book as it was progressing. And José Martini Costa was an excellent research assistant during critical parts of the writing process.

I presented parts of the book at Institut Barcelona d'Estudis Internacionals, London School of Economics, Organization for Economic Cooperation and Development, University of Denver, University of Pennsylvania, the San Francisco Federal Reserve, and Sciences Po Paris. I received fantastic comments and tough questions. Other colleagues, including Rasmus Corlin Christensen, Jeff Colgan, Nick Godfrey, Fergus Green, Rebecca Kysar, Heather MacLean, Jonas Meckling, Matto Mildenberger, Mat Paterson, Tom Peters, Daniel Posen, Lauge Poulsen, Bryce Rudyk, and Dustin Tingley patiently answered questions on a variety of topics ranging from international tax law to life cycle analysis.

A special thanks to Pierre Charbonnier, Danny Cullenward, Jonas Nahm, and David Victor for formative conversations, detailed discussions, and extensive comments on various drafts. The manuscript also benefited from comments on specific chapters by Derik Broekhoeff, Jeff Colgan, Madison Condon, Michael Gillenwater, Tom Hale, Jonas Nahm, Kyla Tienhaara, Todd Tucker, and Wendy Wong.

This is the second time I have had the privilege to work with the crack team at Princeton University Press. Bridget Flannery-McCoy supported this project from the outset and provided excellent feedback along the way. She helped wrangle my many ideas into what I hope is now a much more tractable narrative. As with my last book, Eric Crahan seamlessly took the reins to shepherd the book through its final stages. And Alena Chekanov capably supported the entire process.

Writing a book requires moral support in frequent and liberal doses. I was fortunate enough to have both. Leah Stokes was my writing buddy, and our daily Zoom sessions in the home stretch of editing helped me get the book over the finish line. The Philly crew—Kat, Genny, and Helen—buoyed my spirits and provided friendship, giggles, and emotional nourishment during my visits. My beach buddies Johnny, Crissie, and Alex have provided an oasis of healing, calm, and sanity every summer for the last five years. And my bestie Wendy Wong has seen me through the highs and lows of this book, and all of the life that happened while writing it.

My parents have been the most diehard and consistent cheering squad anyone could ever want. I am lucky to have them on my team. And my sister Julia has, and continues to be, a hilarious and amazing human and source of support. Both Wendy and Julia have generously shared their wisdom, humor, grace, and (sometimes merciless) editorial feedback. I could not have written this book without them.

Finally—and most importantly—in the five years that this book was incubating, my amazing children, Milo and Hannah, have transformed from little kids to fully fledged and completely awesome humans. They are the light of my life and have been an unending source of joy, wonder, humility, laughter, and love. This book is for them.

ABBREVIATIONS

BIT — Bilateral Investment Treaty
CCS — Carbon capture and storage
CDM — Clean Development Mechanism
COP — Conference of the Parties
CORSIA — Carbon Offsetting and Reduction Scheme for International Aviation
ECT — Energy Charter Treaty
ESG — Environmental, social, and governance
ETS — Emissions trading scheme
EU — European Union
GHG — Greenhouse gas
ICAO — International Civil Aviation Organization
ICROA — International Carbon Reduction and Offset Alliance
ICSID — International Center for Settlement of Investment Disputes
IEA — International Energy Agency (IEA)
IO — international organization
IPCC — Intergovernmental Panel on Climate Change
IRA — Inflation Reduction Act
ISDS — Investor-State Dispute Settlement
MNC — multinational corporation
MSR — Market Stability Reserve
NDC — Nationally Determined Contribution
NGO — nongovernmental organization
NZ — net zero
OECD — Organization for Economic Cooperation and Development
PACM — Paris Agreement Crediting Mechanism
UNFCCC — United Nations Framework Convention on Climate Change
USMCA — United States–Mexico–Canada Agreement
VCM — voluntary carbon market
WTO — World Trade Organization

PART I

Shifting the Mental Model

1

We Are Not All in This Together

THE CLIMATE CRISIS is evidence of an incredible governance failure. The year 2025 marks three decades of global climate negotiations. The first annual Conference of the Parties (COP)—where government delegates negotiate international climate rules—was a small affair held in Berlin. Roughly nine hundred delegates attended, along with a smattering of climate-focused nongovernmental organizations (NGOs). Two decades later, when the landmark Paris Agreement was signed in 2015, there were more than seventeen thousand state negotiators and eight thousand nonstate actors in attendance. By 2023, the number of government delegates attending the negotiations ballooned to almost forty-four thousand people, with some eighty-five thousand attendees in total.[1]

It is therefore not an exaggeration to say that millions of working hours have been devoted to multilateral climate cooperation. And yet the climate crisis continues to accelerate. June 2024 was the twelfth consecutive month with global average temperatures 1.5 degrees Celsius above pre–industrial era levels.[2] That figure will likely be higher by the time you read this sentence. Countries closer to the poles are even warmer. Canada, where I live, is now twice as warm as the global average.

The alarm bells are growing louder, and the climate disasters deadlier and more frequent. But efforts to take action as prescribed by the Paris Agreement, the 2016 international treaty that set the goal of limiting warming to 1.5° Celsius, continue at a plodding pace. Roughly three-quarters of all nations have made individual pledges to reach "net zero" emissions by midcentury.[3] Yet all the evidence indicates that we are set to blow through the Paris temperature target—or may have done so already.[4]

Using the lens of existential politics, this book explains why global climate governance—which includes the UN Framework Convention on Climate

Change (UNFCCC), the Paris Agreement, and a host of voluntary efforts—is failing. It also explains why global climate governance will continue to fail unless we shift our understanding of the nature of climate politics.

Existential politics offers a different way to think about climate change—as a political contest between different kinds of asset owners.[5] Climate policy to curb emissions and phase out fossil fuels will lead to trillions of dollars in stranded assets, creating clear winners and losers.[6] Those with large endowments of fossil assets will be the losers. Owners of green assets will be the winners—as well as the basis of a decarbonized economy. But so far, green asset owners are few in number and weak compared to fossil asset owners. For decades, the losers have been running the show, obstructing climate policy to preserve the value of their assets.

The revaluation of assets—through both climate change and climate policy—creates existential politics. In the most extreme cases, people will lose their lives and places will be wiped off the map. Assets will become worthless and firms and industries will collapse. These are the highest possible stakes. The ensuing conflicts are the stuff of existential politics.

But to date, governments have failed to confront these realities. Instead, they have been almost obsessively focused on emissions—measuring, reporting, and buying and selling them. This technocratic approach has yielded some reductions, but not nearly enough to avoid the worst effects of climate change. Worse still, not only has measuring emissions created the illusion that we are all in this together, but it encourages us to believe we are actually tackling the climate crisis—when all the evidence indicates the opposite. Global climate governance isn't working because it is overly focused on the wrong problem.

This book takes a different view of climate change—as a problem of assets, not tons of greenhouse gas emissions. I show that global policies to manage emissions are both failing to promote the energy transition and providing political cover for maintaining the status quo. Using global financial institutions as the central tools for climate governance can meet the twin challenges of constraining the power of fossil asset owners and expanding the number of green asset owners.

Climate change is about loss and transformation. This period of intensifying climate crisis is a deeply unsettling time, steeped in fear of certain change with uncertain distribution. But with a reorientation around existential politics, governments can make meaningful progress on decarbonization.

A Simple Model of Climate Politics

To understand how asset revaluation shapes global climate politics, I conceptualize a simple world with three groups of asset owners: fossil, green, and vulnerable. Firms and governments are the primary asset owners in this model. These collective actors with organized interests are the engine of politics. Of course, individuals can own assets too, but unless they are organized into a group (a union, stockholders, pension owners), they do not have a collective set of preferences or interests. Asset owners' interests are driven primarily, but not exclusively by material concerns, though these interests may change over time as technology advances[7] or ideologies shift.[8]

First are *fossil asset owners*. Fossil assets are currently the engines of the global economy. They exacerbate climate change and, barring major transformations, their owners will lose out from the energy transition. Fossil asset owners include fossil fuel and petrochemical companies, utilities, and heavy manufacturers, among others. Of course, at some level all asset owners hold some fossil assets, since fossil fuels continue to be the lifeblood of our world. Some fossil assets are highly specific—that is, they cannot be replaced, either due to a lack of technological substitutes or prohibitively high costs. But others are potentially "decarbonizable": technological development and diffusion have reached a point where fossil assets can be converted to green assets, given the right incentives.[9] Targeting decarbonizable assets is a critical political leverage point in existential politics.

Fossil asset owners' primary interest is, unsurprisingly, maintaining the status quo, either through outright obstruction on climate policy or by slowing the pace of the energy transition through strategies such as hedging or greenwashing. Indeed, some fossil asset owners, like fossil fuel companies, already have a long and well-established record of doing so. We now know that "Exxon knew" decades ago about the effects of greenhouse gas (GHG) emissions on the climate.[10] And while climate denialism is no longer mainstream, there is ample evidence of the fossil fuel industry and other fossil asset owners seeking to slow the pace of decarbonization through lobbying,[11] self-regulation,[12] and greenwashing.[13]

Even at its current middling pace, climate policy is already devaluing fossil assets. Energy companies are selling off their most carbon-intensive holdings,[14] pensions are divesting from fossil fuels,[15] and some banks now require decarbonization plans to "derisk" their lending portfolios.[16] And some fossil asset owners are coming to terms with the revaluation already underway and beginning to transition to clean technologies. The auto industry is an excellent example. With a relatively mature technology—battery-powered electric vehicles (BEVs)—many major auto manufacturers have ramped up their production. But this choice is far from universal. Automakers such as Fiat and Ford are lagging behind peer firms in their BEV production.[17] Others, like Toyota, continue to bank on hybrid vehicles, despite broader trends toward full electrification.[18]

But of course, not all industries have developed to the point where decarbonizing is technologically feasible. For instance, there is currently no way to electrify aviation. Sustainable aviation fuel (fuel derived from biomass) can reduce, but not eliminate, GHG emissions. These fossil asset owners are more likely to organize and obstruct decarbonization efforts. Thus, the aviation industry's strategy has been to focus on massive offsetting efforts. In 2016, governments created an international agreement to regulate aviation emissions. In the short term, most reductions are expected to come from offsetting. But as I explain in chapter 4, offsetting is a hugely problematic policy, the benefits of which have been systematically and grossly overestimated.[19]

In other industries, technologies are available, but not scalable. The fossil fuel and electricity industries have pinned their hopes on carbon capture and storage (CCS). However, except for storage at the site of combustion, such as in power plants, this technology is not yet economically viable.[20] Some research suggests that even in this limited application, CCS has logged many more failures than successes.[21]

Second are *vulnerable asset owners*. The effects of climate change are creating tremendous losses—of lives, homes, regions, industries, and eventually, perhaps, entire nations. Vulnerable asset owners are those who will lose as climate change intensifies. They include home- and landowners in low-lying areas, the insurance industry, and farmers and fisherfolk, among many others.

Wildfires and floods have decimated communities around the globe, creating huge financial losses and, more importantly, displacing former residents. The International Organization for Migration estimates that there will be over 200 million climate migrants by midcentury.[22] Currently, climate migrants have no legal status under international law, and there are limited plans for dealing with massive flows of people. Whole nations will be erased by sea-level rise, and some are already making relocation plans.[23] Australia recently signed an agreement with Tuvalu, agreeing to accept climate migrants as the tiny island nation shrinks.[24]

The radical uncertainty surrounding climate change is also posing existential threats to the insurance industry.[25] A recent study estimates that extreme weather events linked to climate change created economic losses of US$2.86 *trillion* in the last two decades, or an average $143 billion per year—roughly the annual GDP of Morocco.[26] In Canada, insurance claims around disasters like fires and floods are up more than 400 percent over the last fifteen years.[27] In 2022, there were CAD$3.4 billion in catastrophic insurable losses.[28] In the United States, state-level insurance plans are facing the twin pressures of low premiums and increased payouts. These programs are insurers of last resort, often offering protection to homeowners in fire- or flood-prone areas who cannot get private insurance. Some programs are facing the real threat of insolvency.[29]

Vulnerable owners are also fighting for their survival—but through aggressive mitigation and adaptation policies. It is not surprising that Tuvalu is part of a bloc of small island nations that have long been a loud voice at the climate negotiations, calling for more ambitious policy as a last-ditch effort to avoid the sea-level rise that will submerge them entirely. Funding for adaption as well as compensation for "loss and damage" are both key political issues for vulnerable asset owners. In the best-case scenario, they can implement measures to enhance resilience in the face of climate change; in the worst-case scenario, they can be compensated for irreversible economic and non-economic damage. Yet, in general, vulnerable asset owners are less powerful and not as well organized and therefore face hurdles in effectively pressuring governments to act.

Third are *green asset owners*. They will be the basis of the decarbonized economy. Green asset owners are those involved in the extraction and production of critical minerals, the production of bulk materials in renewable technologies (for example, steel, cement, and aluminum), and the manufacturing of renewable and green technologies (for example, solar panels, wind turbines, heat pumps, and batteries.) Green asset owners also include infrastructure owners, such as electric utilities, which will benefit from expanded electricity grids, charging stations, and retrofitted buildings. With respect to infrastructure, governments are also green asset owners. Thus far, green asset owners are fewer in number and decidedly less powerful than fossil asset owners—a key problem that I discuss later.

The creation of green assets will necessitate huge amounts of labor—to manufacture, assemble, install, and maintain new green technologies. The number of jobs in the renewable energy industry almost doubled between 2012 and 2022 and now sits at approximately fourteen million.[30] Currently,

China leads the world in renewable energy employment, representing 41 percent of total jobs globally.[31] While existential politics focuses primarily on asset owners, the demand for labor can be an opportunity to create a broad base for support on climate action, involving both the transitioning of fossil asset labor and the addition of new green asset labor.[32]

Like all models, this model of asset owners simplifies reality in order to make generalizations; in the real world, all three categories are more complex.[33] Most owners hold a mix of assets, and therefore their interests fall on a spectrum. Their position is determined by the relative proportion of each set of assets as well as the economic and technological ease with which they can substitute assets. I discuss the challenges of messy boundaries between categories in further detail below and also in chapter 2.

Asset Revaluation Drives Existential Politics

Both climate change and climate policy will generate existential politics: the increasingly contentious political battles over which assets, professions, cultures, and nation-states will survive. Of course, all politics is distributional: it is about who gets what, when, and how. Existential politics magnifies these struggles; indeed, it is distributional politics on steroids.[34]

In distributional politics, actors may win or lose things of greater or lesser value. Increased energy taxes or new technology standards will raise costs for energy producers and consumers and could make exports less competitive. Such policies can negatively affect fossil asset owners' bottom line.

Existential politics is a subset of run-of-the-mill distributional politics, which involves: "(1) something of central importance to a given actor being at stake and (2) the prospect of its total elimination."[35] In contrast to distributional politics, substitutes are unevenly unavailable. This means the destruction or complete devaluation of assets, which effectively determines whose way of life gets to survive.

Full decarbonization will mean an end to fossil fuel extraction, a complete devaluation of oil and gas reserves, and the phasing out of all fossil fuel–based technologies.[36] Unless governments and firms involved in these activities develop an equally profitable carbon-free business model—assuming the technology is available to do so—their assets will lose all value. One study estimates there will be over \$1 trillion in stranded assets under a midrange scenario, consistent with 3.5 degrees Celsius of median warming in the twenty-first century.[37] Thus far, no major fossil asset owners—either governments or firms—have publicly committed to a phaseout of fossil fuels, despite the fact that many have simultaneously pledged to go net zero in the coming decades—balancing emissions with removals.[38]

Fossil asset owners can respond to asset revaluation in different ways. In addition to obstructing climate policy, they can greenwash—devoting resources to looking climate-friendly rather than actually behaving that way. They can hedge—investing in both green and fossil assets without fundamentally changing their business model.[39] They can divest—selling off their dirtiest assets to new fossil asset owners. Or, if the technology is available *and there are sufficient incentives to do so*, they can convert their fossil assets into green ones.

The only two strategies relevant for existential politics are obstruction and conversion. Obstructionism is why we have failed to make progress on the climate crisis—despite more than three decades of diplomacy and international rulemaking. Conversion will turn fossil asset owners into green asset owners, changing both their emissions profile and their interests in decarbonization.[40]

How Existential Politics Explains Climate Failures

Existential politics helps explain the persistent failures of global climate governance. There has been a mismatch in strategies by governments and fossil asset owners.[41] Some fossil asset owners have seen the existential threat of climate regulation from the earliest days of the climate regime. Their playbook has therefore *always* been obstructionism.[42] Governments, facing the difficult task of overcoming this obstructionism, have diligently ignored it, until recently. Instead, they have remained stubbornly focused on emissions. In particular, global climate policy has been almost singularly fixated on measuring, reporting, and trading tons of GHG emissions—which I refer to as "managing tons." Since managing tons resolutely ignores the underlying conflicts created by asset revaluation, it cannot, by definition, achieve the rapid emissions reductions that addressing the climate crisis requires.

Obstructionists Are the Driving Force in the Global Climate Regime

"Political will," which will allow governments to enact the ambitious decarbonization measures the climate crisis requires, is often invoked as the elusive missing piece in climate policy. Many factors contribute to a lack of political will, but the political power of fossil asset owners is the elephant in the room. They are extremely well resourced and well organized.[43] The fossil fuel industry's decades-long efforts to undermine the science of climate change is testimony to their early understanding of existential politics.[44] They quickly recognized that real efforts to combat climate change would mean the end of their industry.

But the fossil fuel industry is not the only bad guy. Electric utilities have also tried to slow the pace of the energy transition. As momentum around US renewable energy laws increased, "electric utilities realized these laws could . . . threaten their assets," and they organized to "block, weaken or rollback climate policies,"[45] with a particular focus on rooftop solar.

The animal agriculture industry, which is responsible for an estimated 14.5 percent of global emissions, has also lobbied actively against US climate policy.[46] Meat and dairy production emit vast quantities of methane and contribute to land use conversion. Consistent with the existential politics framework, they are sending lobbyists to the COP negotiations in increasing numbers.[47] Even the auto industry, a potentially "convertible" sector, has fought against fuel efficiency standards,[48] and some manufacturers continue to insist on a future for internal combustion engine vehicles, despite the fact that more than twenty countries have announced plans to phase out their sale in the coming decades.[49]

Obstructionism is not limited to the United States. For instance, South Africa is largely coal-powered. Eighty-three percent of the country's total emissions and 70 percent of its electricity are derived from coal.[50] There are strong links between policymakers and the coal industry, and as such, domestic policy continues to underwrite the coal industry in several ways.[51] Both supply and production have grown since 2000,[52] and coal remains untaxed when used for transport, heating, or process purposes. The carbon tax remains low (around $10 per ton)[53] and contains many exemptions.[54] Australia has a similar story; the coal lobby has been instrumental in slowing decarbonization efforts.[55]

Unfortunately for the climate, vulnerable and green asset owners have nowhere near the same power over climate politics. Some vulnerable asset owners are fighting rearguard actions, such as suing fossil asset owners for climate damages. In a landmark case in the Netherlands, several environmental NGOs sued Royal Dutch Shell to force it to reduce its emissions. The Dutch court ruled in favor of the NGOs, requiring Shell to reduce emissions by at least 45 percent from 2019 levels by the end of 2030.[56] In 2024, Shell appealed the decision—a strategy that is textbook obstructionism—and won the appeal.[57]

Similarly, a handful of California towns have sued fossil fuel companies for the effects of sea-level rise on their homes.[58] And Indonesian nationals are suing the construction company Holcim for climate change–related damages and a drastic reduction in their emissions.[59] In the best-case scenario for climate ambition, these defensive actions can help reduce the material power of fossil asset owners through damages awarded. But they are also piecemeal, slow, and resource-intensive. Most importantly, these actions will have little effect on fossil asset owners' incentives to decarbonize.

Vulnerable asset owners are heterogenous—geographically dispersed, with a variety of interests. Many of them, unsurprisingly, require governmental support for adaptive measures so that they can protect their assets from the worsening effects of climate change. Yet the politics of adaptation is fundamentally different from mitigation; it is reactive and highly uneven.[60] With the exception of a few powerful industries—notably agriculture and insurance—vulnerable asset owners cannot be expected to counter the power of fossil asset owners.

Green asset owners are similarly limited in their influence. Trade associations representing fossil asset owners are typically opposed to climate policies, and they are outspending renewable trade associations by a factor of fourteen to one.[61] Yet it is estimated that the renewables industry could generate up to twenty-four million jobs and increase global GDP by 1.1 percent ($1.3 trillion) by 2030 if governments continue to invest in the energy transition.[62]

In sum, *obstructionism is the key driver of climate politics.* For some fossil asset owners, asset revaluation threatens complete extinction in the face of stringent climate policy; the luckier ones will be able to rebalance their holdings and retool their growth strategies. Green asset owners could serve as a counterweight to fossil asset owners' obstructionism—*if governments invest in their expansion.* But as I elaborate later, these investments will not be made through the UNFCCC. Rather, building a decarbonized economy will require turning our collective focus to the rules of international trade and finance institutions.

Managing Tons: The False Hope of Cooperating with Obstructionists

Existential politics lays bare the reasons that fossil asset owners will obstruct progress on decarbonization to maintain the value of their assets. Yet governments doggedly insist upon cooperation with these same obstructionists through the UNFCCC and the Paris Agreement. Thus, there is a profound mismatch between current approaches and political reality, based in the false hope that multilateral cooperation on emissions reductions can adequately address the climate crisis.

This false hope has consistently taken a particular form. Instead of viewing climate change as a problem of existential politics, governments have fixated on "managing tons." Carbon pricing, carbon offsets (also referred to as "carbon credits"), and net zero policy pledges are all signature policies of the climate regime and prime examples of this technocratic management approach.

Managing tons assumes that measurement and commodification of emissions will lead to climate solutions. This approach is consistent with many other contemporary approaches to global governance that emphasize process over outcomes and governance by indicators.[63] However, this mechanistic mode of governance buries politics in numbers. Policy becomes technical, focused on processes of measurement, reporting, and evaluation.[64]

I am not the first to observe the obsessive focus on emissions in climate policy. Geographer Eric Swyndegouw describes CO_2 as having become "the 'thing' around which our environmental dreams, aspirations, contestations as well as policies crystallize." He describes a "fetishist disavowal of the multiple and complex relations" that drive climate change, resulting in "reductionism to this singular socio-chemical component (CO_2)."[65] Similarly, Larry Lohmann has noted that the highly technical nature of carbon markets effectively suppresses "public discussion [and] is precisely the opposite of the wide-ranging grassroots debate and political mobilization that the climate crisis calls for."[66]

Managing tons is a maladaptive political coping mechanism that has grown out of the constraints of multilateralism. The climate regime requires consensus for decision-making. Since any government can (at least in principle) exercise veto power, decisions often represent the lowest common denominator—the preferences of the least ambitious nation or nations.[67]

Because cooperation is the goal, governments are forced to focus on areas of agreement to make progress. This means that they naturally downplay their differences—the distribution of winners and losers that asset revaluation creates. Areas of agreement are consistently the policies that manage tons.

Managing tons is maladaptive because it allows the climate regime to hobble along, fastidiously ignoring the fundamental conflicts that asset revaluation creates. Instead, it displaces these political conflicts onto technical debates about measurement and commodification. Although this is undoubtedly a practical strategy for promoting cooperation, managing tons has had a limited effect on emissions reductions. It allows obstructionists to redirect efforts toward counting emissions rather than reducing them.

Existential politics explains why managing tons cannot produce the transformations needed to decarbonize the economy. The power asymmetries between fossil and green asset owners effectively enable fossil asset owners to capture policies that manage tons—carbon pricing, carbon offsets, and net zero. However, such capture can be difficult to detect because of the technical complexity (and therefore opacity) of these policies and the veneer of legitimacy imbued by multilateral cooperation. The incredible difficulty of measuring many types of emissions provides ample opportunities for gaming and greenwashing, as I show in chapters 3, 4, and 5.

Managing tons favors fossil asset owners. It also fails to acknowledge the fundamentally conflictual relationship among asset owners, instead insisting that those who stand to lose the most from decarbonization will be proactive contributors to the process. This approach creates a specific form of regulatory capture in which everyone agrees to implement highly technical policies that are often difficult to understand fully and therefore easily subject to gaming and manipulation.

Global Climate Governance Should Focus on Assets, Not Tons

Existential politics dictates a very different path for the global climate regime than managing tons: multilateral action that lessens the power asymmetry between fossil and green asset owners. Policies that manage tons are shrouded in the technical complexities of GHG measurement, offer intangible and often far-off benefits, and are too easily twisted to serve the interests of fossil asset owners. Moreover, concentrating on tons often marginally increases efficiency in a fundamentally fossil-fuel based system. This incremental approach is unlikely to lead to complete decarbonization, which requires wholesale transformations of economic, social, and technological systems.[68] The reality is that we must stop burning fossil fuels.

Ironically, global climate policy has stagnated because governments are too focused on global climate policy. To get to the root of the climate crisis, states should turn to global rules that reorient the flow of capital in the global economy. Reducing the supply of fossil fuels should follow. This approach can constrain the power of fossil asset owners *and* build green assets, thus expanding political support for decarbonizing the economy. *Existential politics indicates that global economic institutions, rather than the UNFCCC, must be the central locus for global climate policy.*

Economic institutions respond directly to the material struggles that existential politics creates. Current trade and finance rules protect fossil asset owners and make it more difficult for governments to invest in green asset owners. These rules cement—and potentially widen—the power asymmetry between these two groups of asset owners, tipping the balance further in favor of obstructionism.

Reform of international economic institutions could help shrink this gap by constraining the material wealth of fossil asset owners and allowing domestic governments the legal leeway to invest in green asset owners. (Because of the diversity of vulnerable owners, they are unlikely to serve as a counterweight to fossil asset owners at the international level, as I discuss further in chapter 2.) Specifically, tax and finance institutions can constrain fossil asset owners by reducing their wealth and political influence. And trade institutions can facilitate the creation of green assets, funneling investments toward the goods, services, and labor needed to build and maintain green assets. One set of reforms promotes the decline of fossil fuels; the other promulgates low-carbon alternatives.[69]

Tax reform can be powerful climate policy.[70] The globalization of the financial industry has made it much easier for corporations to avoid taxation by moving profits offshore to tax havens.[71] Offshoring not only exacerbates wealth inequality (a key cause of climate change) but also builds the wealth of global corporations that contribute to climate change, both directly and indirectly. Directly, companies that offshore are linked to tropical deforestation.[72] Indirectly, corporate offshoring increases wealth inequality, which contributes to climate change.[73] Fossil asset owners, such as oil and gas companies, are also active in the offshoring game.[74] Raising corporate minimum taxes—a process that is already underway via the Organization for Economic Cooperation and Development (OECD)—can help rein in the power of fossil assets.

Curtailing investment protections for fossil asset owners is another important way in which concentrating on assets can accelerate decarbonization. Since 1980, states have signed over 2,600 international investment treaties.[75] Conflicts over the agreements are adjudicated through the Investor-State Dispute Settlement (ISDS) system—the arbitration provisions found in almost all investment treaties. ISDS provisions allow foreign investors to sue states for compensation if domestic regulations impede their investments. For example, the UK oil and gas firm Rockhopper Exploration sued the Italian government for losses when the latter banned offshore oil and gas drilling within twelve nautical miles of the coast. Rockhopper had previously been granted concessions to extract oil from the Ombrina Mare oil field, but the subsequent ban rendered this no longer possible. The ISDS ruled in favor of Rockhopper, awarding €184 million in damages.[76] Similarly, the Canadian government sought $15 billion in compensation when the United States canceled the Keystone XL Pipeline.[77] The pipeline was a flashpoint in North American climate politics, provoking vocal public opposition because the project would have added as much as 110 million tons of CO_2 emissions annually.[78] The case was dismissed in July 2024.

Worse still, protections afforded by the ISDS have resulted in massive payouts to fossil asset owners, and to the fossil fuel industry in particular. Just a handful of lawsuits brought by oil and gas majors have resulted in state payouts of over *$67 trillion* since 2013.[79] Such payouts embolden firms faced with asset revaluation, reinforce their power through wealth accumulation, and may dissuade states from implementing aggressive climate policy for fear of legal reprisals from firms whose investments are protected by the ISDS.[80]

Finally, the turn toward green industrial policy—the strategies governments employ to expand climate-friendly economic activities—shows that existential politics is germane to understanding global climate governance. The recent passage of the US Inflation Reduction Act and the European Green Deal illustrates the political popularity of domestic investments in green assets even in countries, like the United States, that have lagged behind on climate policy.

Green industrial policy is, in many ways, the opposite of managing tons. It delivers near-term benefits to interest groups and consumers alike. For example, the 2022 US CHIPS and Science Act provided $53 billion to incentivize domestic chip manufacturing.[81] The US Inflation Reduction Act has delivered almost $500 billion in climate-related investments since its passage in 2022.[82] The European Green Deal, which aims to make Europe net zero by 2050, is funded by half a trillion euros from the EU budget (and another half-trillion euros from co-financing and other private sources).[83]

There is a growing political impulse to source green assets domestically, creating the twin benefits of increased domestic economic investments and green jobs (with the associated political advantages) and greater self-sufficiency. As Europe and North America struggle with fractious relationships with China,[84] both benefits play well in domestic politics.

But the inextricable relationship between the economic and political challenges of decarbonization makes for a complex balancing act among three competing priorities. First, countries must cooperate to massively ramp up production of green assets. Second, there is an urgent need to create more winners—and therefore supporters—of climate policy. Coalitions of green asset owners and labor are needed to destabilize the entrenched power of fossil asset owners. However, such action can quickly shade into protectionism, which will increase costs and ultimately slow the energy transition. Yet even if it shades into protectionism, green industrial policy can create lasting political benefits. Finally, the demand for green assets is so enormous and urgent that duplicative efforts across nations will be necessary, though they may undercut efficiency. The new frontier and fundamental challenge of climate policy going forward will be to manage the trade-offs between efficiency losses and domestic investments in green assets.[85]

The Scope of the Book

This book proposes an expansive new framework for understanding global climate politics. It explains how international institutions *beyond* the UNFCCC and the Paris Agreement can create the conditions for rapid decarbonization. However, we cannot understand the potential role of these institutions without seeing how domestic politics constrains and is constrained by global institutions. A deep dive into national-level interest group politics is beyond the scope of this book. Instead, I use the model of asset revaluation to make some basic assumptions about the balance of fossil and green asset owners domestically, across nation-states, though some specificity at the national level is necessarily sacrificed.

Importantly, I don't take on the critical problem of petro-states, which are the largest fossil asset owners in the world. Many petro-states have nationalized their oil industries, which are generally less transparent to publics and scholars. And importantly, in these countries the logic of building constituencies is often more complex.

The focus is on assets as the basis for the preferences of asset owners and, to a lesser extent, of labor. This choice is deliberate. Climate change has become an increasingly polarizing issue, and people's attitudes about it are deeply rooted in their political beliefs. Polarization is a tough nut to crack. The model of existential politics seeks to supersede these cleavages through structural changes to the economy. Change the rules first, and the choices available to intransigent groups—voters, fossil asset owners, labor—will follow.

I have made a deliberate decision to think about decarbonization as a political-economic challenge rather than one of justice. I am deeply sympathetic to arguments that climate solutions are intrinsically linked to broader issues about inequality and systems of oppression like colonization and debt. Other scholars have written elegantly and thoughtfully about these relationships.[86]

The "radical pragmatist" solutions set forth in chapters 6 and 7 acknowledge the limitations of my own thinking. The urgency of the climate crisis demands profound changes to the global economy at an incredibly rapid pace. While equity, justice, and decolonization must be part of a politically stable and environmentally sustainable planet, creating a new global order premised on these principles is not likely to come about in the short window of time we are now facing. There are certainly those who argue that a just transition cannot occur in the absence of fundamentally reconfiguring the distribution of wealth and power across and within nations. I acknowledge the importance of these discussions, but do not address them directly in this book.

The Structure of the Book

The book has three main parts. In chapter 2, I present a stylized model of the world with the three sets of asset owners—fossil, green, and vulnerable. It expands upon previous work with my colleagues Jeff Colgan and Thomas Hale by adding the critical group of green asset owners.[87] The chapter explains why managing tons has become the prevailing approach to global climate governance and discusses the different strategies that fossil asset owners adopt. Finally, it explains why focusing on assets makes for better politics as well as better policy.

Chapters 3 through 5 discuss the pernicious politics of managing tons. I argue that managing tons deliberately takes the politics out of climate policy and thus is doomed to remain in the realm of incremental improvements—creating emissions reductions without true decarbonization. While incrementalism might be appropriate for some problems, the logic does not hold for climate change. I provide evidence from three policies that manage tons to demonstrate the deeply problematic nature of this approach. I also show that these policies do little to create green assets.

Chapter 3 provides an analysis and critique of the ur-policy of managing tons: carbon pricing. Once considered the "only game in town" for climate policy, carbon pricing has since been downgraded to "one tool in the toolbox" to address climate change. But the evidence on emissions reductions does not match the rhetoric—or the political costs. To date, carbon pricing has been shown to be an immensely controversial policy in some places, with very limited effects on emissions. Even the European Union, which has the oldest and largest emissions trading system and arguably the most technical capacity to create and administer a carbon market, has achieved only somewhere between 1 and 2 percent reductions per year.[88] Put simply, in many cases, carbon pricing is a political third rail that doesn't produce emissions reductions commensurate with the backlash it generates.

Chapter 4 makes the case for why it's time to get rid of all nonpermanent offsets. After studying offsets for almost two decades, I have seen their profound impact on the logic and politics of global climate governance. Offsets (or carbon credits) were the political linchpin of the Kyoto Protocol that allowed countries from the global North and South to forge an agreement based on the transfer of wealth via offset markets.[89] Private authority has also flourished with the growth of offset markets. Since the late 1990s, nonstate actors, primarily NGOs, have created their own "voluntary" market that makes carbon credits available to those actors who want to offset their emissions but are not subject to regulation requiring them to do so.[90] These self-regulated voluntary markets are rife with quality and integrity issues. Chapter 4 provides a primer on the technical aspects of offset project design and implementation and explains why voluntary markets are especially problematic: they are structurally incapable of solving the quality issues through self-regulation, and their use in compliance markets is growing.

Chapter 5 unpacks the newest organizing principle in climate governance: net zero emissions. Net zero is enshrined in the Paris Agreement, which calls for achieving "a balance between anthropogenic emissions by sources and removals by sinks of greenhouse gases in the second half of this century."[91] Net zero is yet another version of managing tons, one in which actors remain fixated on measuring and reporting their emissions. Over 88 percent of emissions are now covered by a net zero pledge.[92] But the real meaning of these pledges depends highly on the details of the individual pledge. I detail the history of net zero, what currently constitutes best practice, and how current pledges from governments and corporations stack up.

Chapters 6 and 7 are the "solution" chapters in which I offer new strategies for global climate governance informed by existential politics: regulating assets in the global economy rather than tons of greenhouse gases in the atmosphere. I call this approach "radical pragmatism." It is radical in the sense that it challenges some of the policies of the neoliberal international economic order. But it is pragmatic in its acknowledgment that this economic order is not going to be replaced anytime soon. Thus, the challenge is to understand both the policies and the politics that can help accelerate decarbonization. Chapter 6 tackles the difficult question of how to constrain the power of fossil asset owners; chapter 7 offers principles to guide governments in their investments in green asset owners.

Chapter 6 considers how international taxation and investment protections can be reformed to constrain the wealth and power of fossil asset owners. Global tax rules—which are already being reformed—can recoup some of the private profits offshored in tax havens, help fund government investments in green asset owners, and reduce the material power of fossil asset owners. International investment treaties currently protect fossil asset owners, especially the fossil fuel industry, which has reaped enormous monetary rewards from governments through international arbitration. Rolling back these protections is critical for changing fossil asset owners' calculus on the potential profitability of new investments. Best of all, these reforms are already underway (albeit in different ways), so the task is to ensure that robust rules are backed by powerful nations and that fossil asset owners do not succeed in weakening them.

Chapter 7 examines the current missing piece in existential politics: green asset owners. It considers how to grow their number, breadth, and power through the global trade regime. This process is currently unfolding through major green industrial policy initiatives, including the US Inflation Reduction Act (which is currently being rolled back by President Trump) and the European Green Deal. There are clear political incentives to engage in green protectionism that directs investments domestically to develop green assets. Given the urgent need to decarbonize the economy, however, governments must leverage all the efficiency advantages that free trade creates. Chapter 7 provides principles to guide states in making policies that navigate these trade-offs. I explain why policies like the European Union's carbon border adjustment mechanism (CBAM) and the US-EU green steel deal are trending toward protectionism in the name of creating green assets.

Industrial policy is the new frontier of global climate politics—since it directly addresses the fossil asset owner obstructionism as well as the dearth of green asset owners. Unlike policies that manage tons, industrial policy delivers immediate and direct material benefits to build new interest groups of green asset owners.

The final chapter draws out the implications of existential politics for the future of global climate politics. International institutions are sticky; thus, it is unlikely that the UNFCCC and the Paris Agreement are going anywhere. However, we should adjust our expectations for what these institutions can do. They should continue to be a locus of reporting, transparency, and information exchange rather than the forum for creating climate policy. Existential politics dictates that decarbonization will come from transformation of economic structures and incentives, not measuring tons of emissions. To this end, I offer some "harm reduction" measures to mitigate the negative impacts of these policies in the short term.

The unending stream of bad climate news can be overwhelming. Existential politics provides a new perspective on a seemingly insoluble problem. To date, we have been obsessed with measuring emissions, to the detriment of efforts to reduce them. Instead of managing tons, governments need to invest in green assets. Not only will support for green asset owners provide the much-needed technology and infrastructure for a decarbonized economy, but it will create the political support for rapid and ambitious climate action. A focus on assets rather than tons is the best way to manage the climate crisis.

2

How Asset Revaluation
Drives Existential Politics

EXISTENTIAL POLITICS is a new way to think about and explain climate politics. The basic argument is simple: Both climate change and climate policy will revalue assets. Some fossil-based and vulnerable assets will become worthless, while green assets, as the basis of a decarbonized economy, will become increasingly valuable. This revaluation process will create winners and losers who will adopt different strategies to create or maintain the value of their assets. Unfortunately, right now, the losers—fossil asset owners—are running the show. They have sought to obstruct and slow-walk climate policy in a bid to preserve the value of their assets. And thus far, they have been very successful.

Existential politics explains why this is unsurprising: governments have misdiagnosed the political problem of climate change as a collective action problem of emissions reductions among states rather than an asset revaluation problem among firms. This misdiagnosis has resulted in a global climate regime that is largely concentrated on the wrong thing. Governments have created multilateral rules centered on managing emissions rather than directly addressing the conflicts—and particularly the obstructionism—that asset revaluation creates.

In this chapter, I first explain the origins of the misdiagnosis, providing a brief overview of the evolution of the global climate regime.[1] Second, I describe what governments have done instead: papered over fundamentally political problems by engaging in endless technical debates about how to measure, account for, and trade tons of greenhouse gas emissions. I call this narrow focus on emissions a problem of "managing tons." Third, I explain how existential politics provides a better model for understanding climate politics.

Fourth, I describe different types of asset owners and consider the challenging question of how to delineate green and fossil asset owners, and why we should do so.

I present a typology of strategies that fossil asset owners use in the face of asset revaluation: obstructionism, greenwashing, hedging, divestment, and conversion—where fossil asset owners convert to green assets. And finally, I explain why only two of these strategies matter for the global climate regime: obstructionism and conversion. Put simply, they are the only two strategies that affect the balance between green and fossil assets—and thus the political power between asset owners.

Misdiagnosing the Problem

The global climate regime—by which I mean the UN Framework Convention on Climate Change (UNFCCC) and the accompanying Paris Agreement— has focused relentlessly on state cooperation on mitigation as the solution to the climate crisis. Certainly, this single-mindedness is partly due to the nature of multilateralism: international institutions can lower the costs and risks of cooperation and contribute to public goods, like a more stable climate.[2] But another important factor is a misdiagnosis of the climate change primarily as a collective action problem of emissions reductions. For three decades the received wisdom has been that we can address the climate crisis only if states cooperate on mitigation, guided by action under the UNFCCC and its associated agreements.[3] This belief is deliberately ignorant of the distribution of costs of decarbonization—which is where all the politics happens.

The atmosphere doesn't "care" (if it could) where GHG emissions come from; total atmospheric concentrations are the only relevant datum. As such, climate change has routinely been conceptualized as a *global* collective action problem of emissions mitigation, since it has no political boundaries. Such problems require multilateral cooperation; all states must commit to action to avoid the temptation to free-ride on others' efforts. This has been the standard narrative since the early days of the climate regime, and it has had damaging consequences, including an overemphasis on cooperative policies of reducing emissions and a belief that cooperation with large polluting firms can "unlock" solutions to climate change.

The pervasiveness of the view that climate change is a collective action problem cannot be overstated. The first World Climate Conference convened by the World Meteorological Organization in 1979 concluded that "the climates of the countries of the world are interdependent. For this reason . . . there is an urgent need for the development of a common global strategy for a greater understanding and a rational use of climate."[4]

Social scientists, and in particular economists, have reinforced the collective action frame.[5] Economists quickly identified climate change as a basic prisoner's dilemma: collective action yields better outcomes for everyone, but the individual's incentive is to act in her own self-interest:

> It is well known that collective well-being can be increased if all countries cooperate in managing shared environmental resources like the climate . . . but that if this improved situation is attained, every country will earn even higher returns by free-riding.[6]

"Free-riding" here means choosing not to curb emissions, in the hopes that everyone else will do so.

Enforcement is the obvious mechanism to deter free-riding,[7] but one that is notoriously rare in multilateral agreements.[8] Other economists have noted that "global climate change is the ultimate global commons problem. . . . Because of this, a multinational response is required."[9] The 2006 Stern Review—an influential report commissioned by the UK government—also noted that "an effective response to climate change will depend on creating the conditions for international collective action." Other economists, including Nobel Laureates Thomas Schelling and William Nordhaus, have echoed this view.[10]

International relations scholars have also historically taken the position that climate change is a collective action problem of emissions mitigation[11] and a "classic Prisoner's Dilemma . . . in which the option of not cooperating typically is more attractive than cooperation."[12] (A growing literature emphasizes distributional issues, though much of this work is relatively recent and based on comparative politics.[13]) For instance, Thomas Bernauer writes: "Climate change mitigation is a global collective good . . . whose 'production' requires global collective action."[14] Others emphasize free-riding as a key problem in designing a global climate agreement.[15] Nobel Laureate Elinor Ostrom's concurrence with this framing of *global* climate politics is exactly why she instead proposed a "polycentric" solution to attack the problem at smaller scales.[16]

Whether climate change is a collective action problem is not just an academic debate: it has important implications for policy and politics. The fear of free-riding dictated the structure of the Kyoto Protocol, which set precise targets and timetables for developed-country emissions reductions. And the failure of the Kyoto Protocol created the conditions for the more flexible approach embodied by the Paris Agreement, which allows states to set their own goals in "Nationally Determined Contributions."

Managing Tons: The Obsession with Measuring Greenhouse Gases

The insistence on cooperation has led states, firms, and NGOs to target GHG emissions, through a process I call "managing tons"—*the narrow construction of climate change as an issue of emissions mitigation focused on measuring, accounting for, verifying, and trading tons of greenhouse gas emissions.*

Managing tons has been at the core of the climate regime since its inception. The 1992 Framework Convention calls for states to provide national inventories of their greenhouse gas emissions. Subsequently, the Kyoto Protocol created an international market for carbon offsets. And the Paris Agreement sets the goal of reaching global net zero emissions by 2050 and calls on non-state actors to adopt voluntary measures to do the same. These policies and goals all require sophisticated carbon accounting—much of which is highly technical and prone to massaging, if not outright gaming, of the numbers.

Managing tons stems from the misdiagnosis of climate change as a collective action problem of mitigation. If countries must indeed cooperate to reduce emissions, per the collective action frame, then they need to find mutually agreed-upon policies. Managing tons fits the bill, since it displaces underlying political conflicts of mitigation and decarbonization onto technical debates about measurement and commodification.

But technical debates about measurement and accounting have come to replace more substantive—and conflictual—arguments about asymmetries of wealth and power and the obstructionism of fossil asset owners. Managing tons fails to address the profound political-economic conflicts that climate policy and climate change create. It has become a credible, yet largely ineffective, substitute for the *real* decarbonization policies that asset revaluation implies.

There are two large and interrelated problems with managing tons. First and foremost, it has not been an engine for decarbonization. At best, it promotes incremental improvements. Steven Bernstein and Matthew Hoffmann refer to this problem as the "double trap" of decarbonization.[17] In this "fake dynamism," behavioral changes produce emissions reductions but do not promote progress toward true decarbonization. They note that "many interventions aim to improve the efficiency of systems without generating full decarbonization"—such as switching from oil or coal to natural gas as a lower-emission "bridge fuel."[18] Three signature policies of the Paris Agreement—carbon pricing, carbon offsets, and net zero—appear to fall victim to the double trap of making marginal improvements without accelerating the process of deep decarbonization.

Critics might counter that managing tons was, in fact, an appropriate approach in the 1990s, when atmospheric concentrations of CO_2 were between 350 and 370 parts per million and the climate crisis still seemed a distant and uncertain possibility. This critique does not hold, for two reasons. First, even in a world of CO_2 concentration of 370 parts per million (reached in 2001), the reductions required to stabilize (but not reduce) atmospheric concentrations would have been significant. Indeed, the First Assessment Report by the Intergovernmental Panel on Climate Change (IPCC) notes that stabilizing atmospheric concentrations of GHGs at their then-current levels would require "immediate reductions" of "long-lived" gases (i.e., CO_2) by 60 percent. Even the least ambitious scenario in the IPCC's First Assessment (excluding Business as Usual) would have required significant changes: a wholesale shift to natural gas, large efficiency increases, a reversal of deforestation, and a 50 percent reduction of CFC emissions from 1986 by 2040.[19] The Second Assessment Report notes that "the risk of aggregate net damage due to climate change, consideration of risk aversion and the precautionary approach, provide rationales for actions beyond 'no regrets [policies].'"[20] Moreover, it identifies the dangers of inaction: "delaying responses is itself a decision involving costs."[21]

Second, fossil asset owners were implementing strategies of obstructionism in the early 1990s, an indication that existential politics was already in motion. Many of the largest oil and gas companies joined together in the Global Climate Coalition to sow doubt about the science of climate change, lobby governments, and block international efforts at creating a strong global climate treaty. In response to the US publication of its first National Action Plan in 1992, the Global Climate Coalition noted in a press release:

> There is a clear danger in the debate on global climate change for policy makers to rush into action before the scientific community agrees that the proposed actions will actually impact any climate trend. . . . This is certainly true in the debate on climate change, where significant disagreements exist concerning everything from the validity of computer models . . . to the role

of man-made greenhouse gases in altering the dynamics of the natural greenhouse effect."[22]

Approaching climate change as a technocratic problem of managing tons, even in the early 1990s, would have little effect on fossil asset owners who clearly had already created a game plan of obstructionism.

The second problem with managing tons is political. As I show in chapters 3 through 5, the politics of managing tons is particularly malign. Virtually every actor involved—from governments to firms to NGOs—has a vested interest in promoting, maintaining, and expanding policies that manage tons, *even though these policies have had a limited impact on emissions reductions.*

For governments, managing tons can help minimize the visibility of policies at odds with decarbonization. For example, in 2018, Canada adopted a national carbon price.[23] Prime Minister Justin Trudeau pointed to the carbon price as a critical tool in the climate policy tool kit: "We know . . . putting a price on pollution is the most efficient and powerful way to keep 1.5 alive."[24] Yet, at the same time, the federal government bought a natural gas pipeline that was considered too financially risky by private investors and provided billions of dollars to subsidize the fossil fuel industry.[25] The Liberal government continued to emphasize the carbon price as evidence of its climate leadership, despite other policies that actively undermined decarbonization. Similarly, Norway has had a carbon price in place since 1991, well before it became popular. Yet it remains one of the largest exporters of oil, deriving roughly 5 percent of GDP from oil revenue.[26]

National governments also point to net zero pledges as evidence of their climate commitments, though they continue to expand the extraction of fossil fuels. For example, the United Kingdom has pledged to be net zero by 2050 and put this commitment into domestic law.[27] Since the passage of the 2019 law, the United Kingdom has also committed to expanding drilling for offshore oil and gas in the North Sea, citing the need for domestic energy supply. The International Energy Agency (IEA) has emphatically stated that new fossil fuel extraction is incompatible with the Paris Agreement.[28] Yet Prime Minister Rishi Sunak reasserted his "absolute confidence and belief" that the United Kingdom would meet its target.[29]

For firms, managing tons can serve as a useful replacement for climate denialism, which is now well outside of the mainstream. Instead, climate obstructionism is the more common approach taken by fossil asset owners.[30] They lobby rulemakers, slow-walk internal reforms, greenwash, and hedge by promoting policies that focus on tons. For example, fossil fuel companies often support carbon pricing over command-and-control regulations, particularly when designed with low up-front costs, such as free allowances and liberal offset use.[31] They also engage in hedging strategies—making minimal changes in their business behavior to spread the financial risks of emissions reduction strategies.[32]

For international NGOs, managing tons can create political authority. Partnering with multinational corporations (MNCs) on efforts such as enhancing renewable uptake or reducing deforestation generates additional authority, credibility, and access to corporations and governments alike.[33] Managing tons can allow NGOs to establish forms of private authority.[34] For example, in the wake of the Kyoto Protocol, a number of NGOs created their own standards to govern carbon offsets.[35] Over the last fifteen years, these standards have evolved into a large and lucrative voluntary carbon market, valued at $2 billion at its peak in 2022.[36] This market continues to grow and expand, despite numerous studies and reports documenting its dubious offsets and "zombie credits," which do little to actually reduce emissions.[37]

In short, policies that manage tons are politically expedient. Many different types of actors can claim to be doing their part. Everyone is busy, and actors score political victories with voters, shareholders, and other stakeholders. But ultimately, little progress is made on decarbonization. Everyone wins except the climate.

Proper Diagnosis: Climate Politics as Asset Revaluation

Certainly, I am not the first to suggest that climate change is not principally a collective action problem, nor that the multilateral rules to address climate change are inadequate to the task.[38] Many have been banging this drum for years.

Activists from the global South and scholars of climate justice have long emphasized the disparities of both causes and effects of climate change between the developed and developing worlds.[39] One of the earliest formulations of this view came from Anil Agarwal and Sunita Narain, who criticized a report by the World Resources Institute, a Washington-based environmental think tank, as "an excellent example of environmental colonialism . . . [whose] main intention seems to be to blame developing countries for global warming and perpetuate the current global inequality in the use of the earth's environment and its resources."[40]

Others have long emphasized the agonistic nature of climate politics, generally grounding such analyses in Marxist critiques of capitalism.[41] Andreas Malm shows how asset revaluation—exemplified by the shift from water power to steam power—undermined the power of labor and simultaneously created massive new wealth for capital owners.[42] With the subsequent concentration of steam generation in urban centers, employers had access to a much larger labor supply and no longer had to provide workers with additional amenities (such as food and shelter). Workers were easily replaceable, and the deep pool of labor allowed industrialization to proceed rapidly, creating the foundations of the fossil fuel era.

More recently, Matthew Huber has argued that climate change is best understood through the lens of class conflict. Labor, he argues, occupies critical chokepoints in the fossil fuel economy and therefore can be the engine of decarbonization.[43] And activists have put the problem even more starkly: it's capitalism versus the climate.[44]

While I am sympathetic to these types of critiques, this book has a more modest goal: envisioning ways to speed decarbonization under the current structures of capitalism. More than a decade ago, Peter Newell and Matthew Paterson identified this trade-off quite concisely. Addressing the climate crisis "means in effect, either abandoning capitalism, or seeking to find a way for it to grow while gradually replacing coal, oil and gas."[45]

Time is short—too short, I argue—to reinvent global orders. In 2018, the IPCC stated that our best chance to limit warming to 1.5 degrees Celsius is to reach net zero emissions by 2050, with falling global concentrations of GHG beginning in 2030. We are already somewhere between 1.3 and 1.5 degrees warmer, depending on the estimate.[46] It seems extremely unlikely that we will find, create, and implement new organizing structures for the global economy in that time frame.[47] So, for pragmatists like me, we need to find ways to work within the current set of multilateral institutions and global governance systems.

To do this, we must acknowledge the structural constraints of capitalism without negating the power of agents. As such, I explore how governments can reduce the power asymmetries between fossil and green asset owners by reforming the multilateral system (and the capitalist system upon which it is based) rather than remaking the liberal international order. I call this approach "radical pragmatism." Some may be dissatisfied by this approach, since it arguably reproduces many of the same problems of the climate regime in a different set of multilateral institutions. At the end of this chapter, I explain why I think these reforms, which manage assets instead of tons, have a better chance of producing real decarbonization.

Three Types of Asset Owners

I hope I have established why collective action on mitigation is not a helpful way of thinking about climate politics. Instead, I argue, asset revaluation generates *existential politics*—the conflicts that arise among actors when they face the prospect of the creation of new assets or the devaluation or elimination of the assets they currently hold. Actors include firms, organizations, individuals, and governments, among others. And of course, assets vary widely—from homes to capital equipment to the infrastructure of entire nation-states.

To understand how existential politics works, consider a simplified world with three sets of asset owners. *Fossil asset* owners are currently the dominant group—those that own or have long-term investments in fossil fuel–based assets. These owners include but are not limited to fossil fuel and mining companies, electric utilities, automobile manufacturers, and heavy industry such as steel and cement production, which require large amounts of energy and are difficult to decarbonize. The financial industry is also heavily invested in fossil assets.

Green asset owners are the second critical group in this simplified world of existential politics. Green assets will be the foundation of a decarbonized economy, and as I explain later, their owners are the political linchpin to decarbonization. Green asset owners include producers of renewable energy, zero emissions vehicles, zero-carbon buildings, and the infrastructure required to produce all these goods. Green infrastructure will include expanded electricity grids, charging stations, and retrofitted factories, buildings, and homes. As I discuss in chapter 7, the politics and policies of the green transition are oriented toward these assets and how to produce the requisite technologies on the scale needed to decarbonize the economy. Beyond production, however, services will also be an important part of the equation.

Vulnerable asset owners are those whose assets are threatened by climate change. Farmers whose crops are destroyed by droughts or extreme weather are one example. Insurance companies are another important vulnerable asset owner. As the damages from storms, droughts, floods, and wildfires continue to mount, insurance (and reinsurance) companies face increasing financial pressure. Between 1980 and 2023, the European Union estimates, climate change has cost its member states €738 billion, with almost one-quarter of

TABLE 2.1: Types of Climate-Related Asset Owners and Examples

Fossil Asset Owners	Vulnerable Asset Owners	Green Asset Owners
Fossil fuel companies	Homeowners in fire- and flood-prone areas	Renewable energy companies and utilities
Mining and heavy manufacturing companies	Insurance companies	Producers of solar and wind energy components
Electric utilities	Farmers and fishers	Electric vehicle producers
Banks that finance fossil assets	Governments that own climate-vulnerable infrastructure	Firms that install and service green assets

those losses accrued in just the two years between 2021 and 2023.[48] Some of these costs are borne by insurance companies, while local, state, and federal governments—the insurers of last resort—cover other costs. Other vulnerable owners include coastal homeowners (as well as those in flood- or fire-prone areas) and cities, which often own climate-vulnerable infrastructure.

Asset holdings are a critical determinant of political interests. Because asset owners will organize to protect the value of their holdings, we can generally expect certain behavior from each group:

1) Fossil asset owners will oppose stringent climate mitigation policies, since decarbonization threatens the value, and perhaps the very existence, of their assets.
2) Green asset owners will advocate for more ambitious climate policy, since they benefit from the energy transition, which will increase demand for their assets.
3) Vulnerable asset owners will also want more stringent mitigation and adaptation measures as their assets are increasingly devalued by the effects of climate change.

This model is limited to collective or institutional actors, rather than individuals, since the former represent the organized interests that are the engine of politics. Thus, while an individual may have a pension invested in fossil fuels, she is not a significant political actor until her preferences are represented by the institutional investor. In the real world, most actors hold a mix of each type of asset, though fossil and vulnerable assets currently dominate.

In general, asset holdings provide a good first approximation of whether a given actor will be pro- or anti-decarbonization, but of course, there are other forces at work. Norms, culture, and other social processes also influence asset owners' interests. I acknowledge that these factors are important but do not examine them here.

Like all models, this one simplifies reality to explain the basic preferences and behaviors of the actors within it. But reality is more complicated. Most actors hold more than one type of asset. No asset is entirely green, and some fossil assets are more emissions-intensive than others. As such, the boundaries around the asset categories are blurry and therefore can be contested. Asset owners may seek to "drive in both lanes"—claiming assets from each category when doing so is beneficial.[49] I refer to this as "the boundary problem." Another challenge for the categories in this model is that some owners have highly specific assets that are difficult to sell or convert, while others can more easily remove fossil assets from their holdings. I address each of these issues in turn.

The Boundary Problem

Fossil asset owners can adopt various strategies in response to asset revaluation. They can *obstruct* climate policy through lobbying and public relations or disinformation campaigns. They can *greenwash*, using "more resources to promote the organisation as green than are spent to engage in environmentally sound practices."[50] They can *hedge*, devoting some resources to developing green assets.[51] They can *divest*, selling off their fossil assets to another owner; as noted earlier, divesting depends on asset specificity and does not, in fact, solve the fundamental problem of fossil asset ownership. Finally, if cost and technology allow, fossil asset owners can *convert* their fossil assets to green ones.[52]

In existential politics, only two strategies really matter in dealing with fossil asset owners: obstruction and conversion. Obstruction has kept the global climate regime at its current glacial pace. Conversion transforms asset holdings from fossil to green, and owners' interests flip accordingly—from anti- to pro-climate.[53] All the other strategies are a distraction from the core problems of existential politics: building the political power of green asset owners and constraining that of fossil asset owners.

Thus, to address the boundary problem, I focus solely on two groups: those who perennially obstruct climate policy and those whose assets are "convertible"—that is, they can be converted to green assets with the right incentives and regulations.[54]

Convertible assets are in a tricky category of their own, since conversion is not yet technologically or economically feasible in many sectors.[55] However, Nils Kupzok and Jonas Nahm have identified sixteen industries that are "decarbonizable": they currently use fossil fuels but could transition to hydrogen or renewable fuels. For instance, power generators can switch from fossil-based power to renewables. Auto manufacturers can phase out production of internal combustion engines and instead produce electric vehicles. Buildings can produce the electricity they consume. And electricity-intensive manufacturing can source from green power.[56]

As Kupzok and Nahm note, the possibility of conversion creates new political dynamics: "The fossil coalition has begun to fracture as a growing number of carbon-intensive yet decarbonizable industries have joined green groups in demanding a green fiscal expansion."[57] Thus, traditional fossil asset–based industries like agriculture and aviation are beginning to advocate for climate spending to help them expand their green assets. This fracturing *could* provide opportunities for conversion—if governments get the incentives right.

Green assets also suffer from a boundary problem. Because we have yet to decarbonize fully, no green asset is entirely green. For instance, wind turbines may produce clean energy, but they still must be constructed with emissions-intensive steel and aluminum. The same is true for "net zero" buildings: they may generate all the power they consume, but categorizing them as green assets does not account for the emissions generated during their construction. And production of the batteries and magnets that power green assets such as electric vehicles requires mining rare earth elements, whose extraction comes with its own environmental and social impacts.[58] In other words, green assets are only as green as their inputs, and therefore "greenness" varies at different points along the supply chain. Again, this highlights one of the core political challenges of existential politics: creating the economic incentives to convert fossil asset owners into green asset owners to lessen their opposition to climate policy.

And finally, when considering the conceptual model of different asset owners, what is considered "green" is contested. Does the conversion to natural gas with CCS count as green? What about assets that use direct air capture or mine the deep sea for minerals to use in batteries? Again, concentrating on asset owners' strategies helps answer this question. Are these asset owners engaging in obstructionism—arguing, perhaps, for investment in natural gas at the expense of renewables? If the answer is yes, then they cannot be considered green assets, since their political strategies are slowing decarbonization.

Asset Specificity

"Asset specificity" is the extent to which an asset can be sold or converted for other purposes. Owners with relatively low asset specificity can make changes in their holdings with little to no economic loss.[59] When asset specificity is high, exit is not an economically viable option. For example, Saudi Aramco or Chevron would be unable to decarbonize their assets without suffering huge economic losses, including write-downs on oil reserves, refineries, and extraction technology. Full divestment of these assets would be both difficult and costly. (For this reason, some oil and gas companies are pivoting to carbon capture to repurpose some of their capital.) And divestment then leaves that oil company without a business strategy, for what is an oil company without oil? When assets have limited fungibility, owners' economic and therefore political interests are more limited.

By contrast, a firm that can sell off its fossil assets relatively easily and without considerable losses has much greater flexibility in its political strategies. If climate regulations are poised to become extremely costly, the actor can divest from soon-to-be expensive assets in favor of cheaper, cleaner ones. Thus, if a bank sells off its coal investments, it is less likely to be a climate obstructionist. Similarly, if a car manufacturer decides to increase its production of electric vehicles, it may become a greater advocate for decarbonization.

A given actor may be able to change its position from fossil to green asset owner through reconfiguration of its holdings. But unless these assets are retired, this shift does not fundamentally change the political challenge of asset revaluation, since a new owner will now have the vested interest in those fossil assets.[60] For example, the Danish state-owned oil company divested itself of its oil and gas holdings in 2017 and renamed itself Ørsted. The company now sources 91 percent of its energy from renewable sources, with a plan to be completely carbon-neutral by 2040. However, it did not simply write down the billion dollars' worth of oil and gas assets; instead, it sold them to a British chemical company.[61]

How Asset Revaluation Drives Politics

Dividing political interests into types of asset owners allows us to understand the history of global climate politics. It also explains the current inadequacies of the Paris Agreement, as I detail here.

The Current Landscape: Fossil Asset Owners Are Running the Show

For the last three decades, fossil asset owners have invested massive resources in slowing global climate rules. In the late 1980s and early 1990s, just as states began to consider climate change as an issue on the international agenda, fossil fuel companies joined together to form the Global Climate Coalition—an industry association that sought to undermine the science of climate change. Its campaign of disinformation continued through the negotiation of the Kyoto Protocol.[62]

Though climate denialism has since become largely normatively and socially unacceptable (extremists remain, of course), many fossil asset owners have switched tactics, adopting low-cost mitigation measures or greenwashing to appease regulators and critics. Fossil asset owners have been successful in slowing the pace of decarbonization because many are exquisitely wealthy, well organized, and powerful.[63]

The fossil fuel industry is the obvious "bad guy"—in terms of both wealth and its efforts to undermine decarbonization. Investor-owned oil and gas companies are booking record profits. In 2022, BP, Chevron, Equinor, ExxonMobil, Shell, and Total earned $219 billion in profits, prompting the United Kingdom to implement a "windfall tax" on oil and gas firms.[64] Governments are also major fossil asset owners; those with large state-owned oil companies have frequently slowed progress on climate policy. For instance, Adnoc, the state-owned oil company of the United Arab Emirates, plans to invest $150 billion in drilling over the next five years.[65]

Fossil fuel companies aren't the only fossil asset owners engaged in obstructionism. Electric utilities and high-emitting industries like steel are also slowing progress on decarbonization. Leah Stokes has documented how US electric utilities, powered by fossil energy, fought the expansion of renewable energy targets and the net metering policies that paid small-scale producers for distributed solar energy generation.[66] Agriculture, also a big emitter, has also engaged in obstructionism. In Ireland, agriculture generates 37.5 percent of total emissions and employs 7 percent of the population. The Irish Farmers Association, representing major fossil asset owners, is the second most active climate lobbying organization in the country and successfully weakened a recent law to cap the industry's emissions.[67]

While fossil asset owners are wealthy and well organized, vulnerable asset owners face myriad challenges in exercising influence. First, they are geographically diffuse and heterogeneous in that they face very different threats to their assets. In many ways, vulnerable asset owners' primary demand is for adaptation—a policy that spans many ecological and political questions beyond just the question of who will pay. There are issues of social justice, comparative and urban politics, political conflict, security, and migration, among many others.[68] For instance, the challenges for farmers facing droughts are extremely different from those for coastal communities dealing with hurricanes and sea-level rise.

Effective adaptation is highly dependent on financial resources and state capacity. For this reason, distribution of adaptation measures will be highly uneven globally and dependent on international transfers of wealth. Political processes of adaptation will also vary widely. As Robert Keohane notes, "Adaptation perfectly fits what pluralist democracies do best: respond to directly affected concentrated and organized interests with targeted benefits."[69] In autocracies, however, benefits are not likely to be distributed so transparently.[70]

As a result of multiple levels of heterogeneity, vulnerable asset owners often seek remedies in a piecemeal fashion. Farmers may lobby governments for more support in the event of drought, while communities threatened by sea-level rise sue fossil asset owners for causing climate change. Because of the diversity of assets that vulnerable asset owners hold, they are generally not organized at the international level.

The one important exception to this observation is further evidence of the power asymmetry between fossil and vulnerable asset owners. Small island states represent the pinnacle of existential politics: they will literally be wiped off the map by sea-level rise caused by climate change. They have organized into a small but vocal coalition at the climate negotiations, the Alliance of Small Island States (AOSIS). Even before the Framework Convention was signed in 1992, AOSIS was advocating for financial protections from "loss and damage" incurred because of climate change.[71] Yet it took thirty years, until COP27 in 2022, for all nations to agree to establish a fund for loss and damage. While this new fund is undoubtedly an important political victory for AOSIS and other developing nations, it will be successful only if developed countries contribute. Thus far, their record on providing financing through a variety of UNFCCC mechanisms is mediocre at best.[72]

Vulnerable asset owners may be proponents of aggressive climate policy, but often they are simply trying to get help for damages already incurred—as the Loss and Damage fund illustrates. In other words, their preferences may skew more heavily toward adaptation, which is now a critical component of climate policy but does not address the power and influence of fossil asset owners.

Finally, green asset owners are critical to both the technical and political processes of decarbonization, but they have yet to flex their political power in global climate politics. Currently, they are relatively few in number and have modest resources and influence.[73] While optimists note that the price of renewables is falling, or that the number of electric vehicles produced is growing, these changes do not necessarily correspond to an increase in the political power of green asset owners. Investment choices are not made based on prices alone but also on present and future capital investments.[74] Many green technologies are still at their nascent stages and require "patient capital" from the state to underwrite some of the risk and the long time frames associated with less certain investments.[75]

Moreover, because many green assets are still in development or have not been widely deployed, the associated labor force is relatively small. This is an important consideration in the power asymmetry between green and fossil asset owners: those working in fossil industries tend to side with the owners of these assets, resulting in a "double representation" of fossil asset interests.[76] This dynamic does not occur with green labor. I explore the political implications of this imbalance in labor interests in chapter 7.

Is a Focus on Assets Better than Managing Tons?

Thus far, I've made three main arguments. First, climate change politics creates winners and losers through asset revaluation. They then become interest groups that have a stake in whether climate policy proceeds, and if so, at what pace. Second, unless they convert their holdings, fossil asset owners will be losers in the decarbonization process and therefore seek to obstruct ambitious climate policy. And importantly for the dynamics of climate change politics, they are well organized and powerful. Third, green asset owners will be the winners, but they either do not exist, are relatively weak, or still hold enough fossil assets to incline them toward hedging. This creates a power asymmetry that favors the obstructionism of fossil asset owners.

These arguments imply three main avenues for making meaningful progress on decarbonization. At the domestic level, states must provide incentives to convert fossil asset owners so that they will choose to shift their holdings and invest in green assets instead. At the international level, global rules should support domestic investment in green assets and curb the material power of fossil asset owners. This is both good policy and, importantly, good politics. Delivering immediate material benefits to green asset owners can help build the coalitions needed to counter obstructionism and advocate for more stringent climate policy. Critically, it can also provide incentives for fossil asset owners to convert their holdings.

Given the decades-long record of obstructionism by fossil asset owners, it is important to ask whether a focus on assets rather than tons will be a more effective approach to global climate governance. Using trade and finance institutions could simply displace fossil asset owners' influence to another realm of global governance and replicate the challenges of technocrats making complex policies.

Perhaps.

But the politics of creating new assets is hugely different from managing tons in three key ways. First, a focus on assets responds to the overlapping crises—climate change, global inequality, migration, rising populism, and war (among others)—that are challenging the very tenets of the liberal international order.[77]

We are now at an inflection point. The postwar institutions created by the global North are being challenged by the rest of the world. The Washington Consensus has created policies that are no longer politically tenable. Global economic integration has chipped away at democracy and national sovereignty,[78] and the current climate regime is struggling to maintain legitimacy and relevance.[79]

Reform is urgent; without it, some have argued, populism and the erosion of democracy will continue to sweep the globe.[80] Others have put the challenge even more starkly: climate change means that the current international system must transform or collapse.[81]

Reorienting capital flows to create new assets is, in essence, a reaffirmation of sovereignty, with global rules supporting rather than undermining states' ability to invest in their economies. In short, the weaknesses and instability of the liberal international order are precisely why concentrating on assets is much more politically feasible than managing tons: there is already urgent political demand for it that can be leveraged by politicians and by interest groups seeking reform.

Second, shifting incentives for investment and the flow of capital provides immediate material benefits to both asset owners and labor. The importance of material benefits cannot be underestimated. Managing tons frontloads costs and backloads benefits; investing in assets does the reverse. The former is a political loser, whereas the latter can garner broader support. For example, one study finds that Americans, especially people of color, are more likely to support climate policy when it is coupled with other social and economic benefits, such as a guaranteed higher minimum wage.[82] And counterintuitively, a focus on assets can also provide benefits to labor, particularly in fossil-dominated industries, which have traditionally sided with capital.[83]

Third, those investing in green assets are at least slightly less prone to massaging the numbers. Dollars, unlike tons, are an accepted global metric. They do not require methodologies, conversion factors, or counterfactuals. Accounting for dollars is widely understood and practiced; consequently, there is much broader knowledge about how to cheat, and that is good for both cheaters *and* watchdogs. Carbon accounting, by contrast, remains the province of a relatively small number of technocrats and is largely illegible to most people. And as the following three chapters demonstrate, managing tons has become a tool used by fossil asset owners to delay more meaningful action.

Finally, when actors do massage the numbers, the political message is much clearer. It is much more difficult to understand what is wrong when companies buy bogus offsets or fail to report their supply chain emissions than it is when companies fail to pay their taxes, or polluters receive government subsidies. Dollars are simply more politically legible than tons, creating more possibilities for transparency, mobilization, and reform.

PART II
Managing Tons

3

The Limits of Carbon Pricing

IN PAINTING the picture of why managing tons is a flawed approach to global climate policy, I begin with carbon pricing, the ur-policy of managing tons. Carbon pricing reduces political conflicts to a number: the price of a ton of greenhouse gases. When priced "correctly," the economic argument goes, firms will change their behavior, investing in clean technologies and practices, and emissions will fall. Economists have promulgated carbon pricing as the most economically efficient approach to emissions mitigation.[1] But many political scientists tend to be more skeptical. As Barry Rabe notes, though carbon pricing may be a good idea, "carbon pricing policies do not necessarily self-implement and flourish."[2]

Why? Because policy is not politics. Changing the price of fossil fuels, whether through carbon trading schemes or carbon taxes, creates highly visible up-front costs while promising future benefits diffused across the globe. The logic is: pay now, and *maybe* reap the rewards (of a lesser-changed climate) later—along with everyone else. Not exactly a political winner.

But despite the political problems with carbon pricing, it has been a darling of centrist and even center-right governments. Twenty-four percent of global emissions are now covered by a carbon price, and that figure is growing.[3] And yet emissions continue to rise.

This chapter explains where carbon pricing came from and why it continues to expand, despite its many problems, as a preferred policy tool both nationally and internationally. I begin with a very brief history of carbon pricing. I show that it originated to avoid political conflict among governments (in the case of the Kyoto Protocol) and between governments and firms (at the domestic level). Firms generally prefer carbon pricing to command-and-control regulations because it provides them with more leeway and greater possibilities for favorable treatment, consistent with strategies of managing tons. There are multiple design features that can assuage political conflicts. These compromises have resulted in a largely agreeable policy that has had a limited impact on emissions.[4]

Though firms tend to like carbon pricing, many voters do not. Indeed, carbon pricing has been extremely unpopular in some jurisdictions.[5] It is red meat for politicians who want to shield voters from additional costs and the source of loud objections from the public (see, for example, the Yellow Vest protests in France). In my home province of Ontario, Premier Doug Ford challenged the legality of the federal carbon pricing program. The case went all the way to the Canadian Supreme Court, which ruled in favor of the federal government.[6] Though this form of managing tons is not always popular with the public, in the main, it is a much more agreeable approach than "stick"-based policies for both the regulator and the regulated.

I make three arguments about the politics and policy of carbon pricing. First, and most importantly, the evidence indicates that carbon pricing has a limited effect on emissions and does little to drive decarbonization. Even as "just one tool in the toolbox," it is simply not delivering significant reductions, raising the question posed by one scholar more than a decade ago: "Who and what are carbon markets for?"[7]

Its limited effects are a feature, not a bug, of carbon pricing. As with other policies that manage tons, carbon pricing is designed to buy the support of fossil asset owners, even under the most pro-climate conditions. This is why cap-and-trade schemes often provide free permits to big emitters and allow the use of cheap offsets, despite mountains of evidence that many of these offsets do not reduce emissions (see chapter 4).

Second, emissions trading schemes (ETSs) require tremendous governmental capacity to administer and implement. They do not, as many politicians claim, "let the market decide." On the contrary, by creating a currency—the emission allowance—the government is constructing a *new* market, which must be actively managed to function well. I examine two cases most likely to succeed: the European Union Emissions Trading System (EU-ETS) and the California cap-and-trade scheme. Both are advanced economies with extensive regulatory capacity and centralized authorities for overseeing market function. I show that, thus far, both policies have necessitated tremendous regulatory efforts that have only resulted in limited emissions reductions.

Third, even in these cases where "success" is most likely, there are myriad implementation problems, including low prices, ample exemptions, leakage (shifting emitting activities outside the regulatory boundaries), the growing complexity of overlapping instruments, and the countervailing effects of fossil fuel subsidies. None of these problems can be avoided or corrected without active management. In general, policy adjustments do little to improve outcomes—which is why it is unrealistic to expect any real progress on decarbonization through carbon pricing. At best, better-designed carbon pricing policies can reduce the impact of these fundamental flaws but cannot eliminate them altogether.[8]

Finally, I discuss what these challenges mean for the newest—and most likely the largest—expansion of carbon pricing: the European Union's carbon border adjustment mechanism (CBAM). In May 2023, the European Union passed a regulation requiring that EU importers purchase allowances equivalent to the carbon content of the imported good, beginning in 2026. It will apply only to the most carbon-intensive goods (which are at greatest risk of leakage): cement, iron and steel, aluminum, fertilizer, and electricity. I show that the CBAM is simply another version of managing tons and is likely to suffer from the same political and policy problems.

Carbon Pricing Basics

Carbon taxes place a surcharge on fuel or energy use. In emissions trading schemes—which I also refer to as cap-and-trade schemes—the government sets a ceiling or cap on the total amount of allowed emissions. Allowances are distributed to those firms regulated by the scheme, either free of charge or by auction. Each firm then has the right to emit up to its share of allowances. Firms may also trade allowances with each other to meet their individual emission allocations. Those who emit more than their allowance can purchase more; those emitting less can sell their excess supply or bank it for future use.

Carbon taxes and ETSs differ in several respects. First, carbon taxes provide certainty of cost: the price is set by the government. Yet there is no limit on emissions, provided that regulated entities are willing and able to pay the tax. By contrast, ETSs provide certainty of quantity: the cap, set by the government, constitutes the upper limit on emissions. The cost will vary, depending on the scarcity (or abundance) of allowances, which is determined by both the design of the ETS and how it interacts with other climate policies. In practice, the distinction between the two policies is sometimes blurred.[9] For example, an ETS might have a guaranteed floor price, which would make it resemble a tax.

Second, compared with ETSs, carbon taxes are relatively easy to design and administer. Governments have lengthy experience in collecting taxes. ETSs, on the other hand, are quite complex. Governments have to set the cap, which determines both the amount of allowable emissions and the entities that will be regulated by it. Caps are partly informed by science, but they are also a function of anticipated costs.

The scope of the ETS determines which entities are covered by the cap. For example, the European Union's ETS covers about 40 percent of its emissions, including oil refineries, coal, steel and power plants, and high-emitting industries like cement and paper manufacturing as well as intra-EU aviation.[10] The South Korean ETS, by contrast, covers about 70 percent of emissions, including industry, buildings, transportation, waste, and agriculture.[11]

After setting the cap and determining the scope, governments must then decide whether to distribute or auction allowances, or both. While auctions are economically the most efficient, they present political challenges since they impose immediate costs on the regulated entities. Many ETSs begin by distributing some portion of allowances for free, then gradually increase costs.[12] In most cases, high-emitting and trade-exposed sectors receive special treatment. For example, in Canada the largest emitters—such as oil, gas, mining, steel, mineral and chemical producers—are regulated under an output-based pricing system. This system prices carbon against an industry benchmark rather than on a per-ton basis, lowering the overall costs for these emitters. Finally, governments must create a platform for tracking, trading, and retiring those allowances.

The use of offsets is another important design choice. If offsets are permitted, then governments must draft new protocols for offset projects designating which types of projects are permissible. The California Air Resources Board (CARB), which oversees its ETS, has approved six different protocols for offset projects. Because California and Quebec have linked their markets since 2014, each ETS accepts offsets generated in the other market. This can introduce problems if there are issues with offset project quality, as I discuss further in chapter 4.[13]

Governments must also create mechanisms for monitoring and verifying offset project outcomes. These tasks are generally delegated to a third party, which must then be accredited by the government—adding yet another layer to the regulatory landscape.[14]

Finally, most ETSs allow firms to bank and borrow allowances from future compliance periods. In theory, this provides regulated firms with both liquidity and flexibility. In practice, it can contribute to oversupply problems, as I show here with the California case.

A Brief History of Carbon Pricing
in the Global Climate Regime

Carbon pricing has been part of the global climate regime almost since the signing of the Framework Convention in 1992.[15] At the very first Conference of the Parties (COP1 in Berlin), governments agreed to a pilot phase (termed "Activities Implemented Jointly") that allowed countries to cooperate on projects to reduce emissions. Though the phase formally began in 1995, a few projects preceded the decision.[16] Activities Implemented Jointly was the first step toward the creation of a global offset market, which was institutionalized in the Kyoto Protocol.[17]

The Kyoto Protocol is now ancient history, yet it is critical to understanding the current state of carbon pricing. Even at its moment of conception in 1997, Kyoto was a tenuous political agreement. It divided the world into two categories—developed and developing nations. Developed nations were required to reduce their collective emissions to 5 percent below 1990 levels by the end of 2012. No reduction requirements were imposed on developing nations. The United States signed the Protocol, but then did not ratify it, citing the lack of developing country obligations as the reason. The prevailing view of climate change as a global collective action problem required that everyone participate, to avoid free-riding. In the absence of participation from the United States (which was then the largest emitter), the agreement hobbled forward. The Kyoto Protocol's relevance and membership waned, until it was replaced in 2015 with the Paris Agreement.

The Kyoto Protocol created new institutions to facilitate carbon pricing in two forms: offsets (Article 12) and emissions trading (Article 17). Both provisions were critical to securing consensus, especially the offsets provision. But they were also both very new and politically contested policies. As a result, the Protocol did little more than establish the goals of each instrument. The details were left for subsequent negotiations.

Indeed, it would take four years to finalize the rules for the global offset mechanism, called the Clean Development Mechanism (CDM).[18] Similar to Activities Implemented Jointly, the CDM allowed developed countries to offset their emissions by paying for emissions-reducing projects in the developing world.

By many accounts, the new market mechanism was the linchpin to securing consensus.[19] Developed countries now had a much-needed escape hatch: if domestic reductions became too politically onerous or costly, then states could instead pay for mitigation activities in the developing world. Developing countries, for their part, saw the CDM as an important revenue stream in promoting sustainable development, but they refused to commit to mandatory reductions. This was a win-win to solve the prickly divide between North and South.

As I detail in the next chapter, offsets were a great idea in theory—a good political bargain, an economically efficient approach to reductions, and a way to promote clean development in the developing world. The tiny catch is that, in many instances, offsets don't actually reduce emissions. Study after study shows systematic overestimations of their impact on reductions, among many other problems.

Chapter 4 goes into detail about these problems. Two key points are worth highlighting in the broader context of carbon pricing, since offsets are often used in tandem with emissions trading. First, offsets have *always* been subject to serious measurement problems.[20] The problem of determining "additionality"—or the amount of GHG emissions avoided or removed as the result of project financing—requires comparing the project's outcomes to a hypothetical counterfactual that is challenging to estimate: the emissions that would have occurred without the project. And second, offsets were never meant "to reduce net GHG emissions, let alone support their near-total elimination."[21] Rather, they were designed to be an optional and short-term cost containment measure as actors charted a path toward broader emissions reductions.

Emissions trading was the other carbon pricing innovation of the Kyoto Protocol. Again, the treaty text purposefully contained no details about how it would actually function—the strategy pushed through by the president of the COP in order to reach an agreement.[22] In his prescient analysis in 1999, *The Collapse of the Kyoto Protocol*, David Victor refers to this significant omission as "agreement by avoidance": "Diplomats from key countries . . . left Kyoto with the assumption that emissions trading was integral to the pact, yet they deferred until later any effort to settle the rules that would govern the system."[23]

Thus, both major carbon pricing instruments were created as diplomatic salves to a contentious negotiating process. Kyoto contained minimal provisions for actually designing and implementing carbon pricing. And as it turned out, both instruments are rife with challenges.

Yet countries have ensured that international markets for emissions reductions and offsets will continue and expand. The Paris Agreement created two new carbon pricing mechanisms. One is an updated version of the CDM to generate carbon offsets. The other allows for countries to bilaterally agree to exchange funding for emissions reductions. I discuss these later in the chapter, where I examine the global expansion of carbon pricing mechanisms.

Does Carbon Pricing Reduce Emissions?

The key question, of course, is whether this form of managing tons results in fewer emissions. On balance, it appears that carbon pricing, and ETSs in particular, have a limited effect on reductions. In 2021, I reviewed all the peer-reviewed literature at the time that had conducted ex-post quantitative examinations of the effects of carbon pricing policies.[24] The findings were unexpected.

First, despite the myriad assertions by policymakers that carbon pricing is critical for tackling climate change, we have astonishingly few studies assessing their performance. I found that among the literally thousands of articles about carbon pricing schemes, only thirty-seven evaluated their ex-post performance on reducing emissions. Many ran models about the *anticipated* effects—on emissions, price elasticity, and employment, among other areas. But very few had examined *actual* performance.

Moreover, there is a huge geographical bias in these few studies: 70 percent of them examine the EU-ETS. On the one hand, this is logical, given that it is the oldest and largest ETS (though this is changing, given the recent growth of emissions trading in China). On the other hand, we know virtually nothing about the performance of carbon pricing in other jurisdictions. Given the broad and oft-repeated claim that we need carbon pricing to tackle the climate crisis, it is worrisome, to say the least, that we have little evidence to support this claim. This point cannot be understated.

Second, and more worrisome still, were the aggregated findings from these thirty-seven studies. Though there was considerable variation across sectors, the overall reductions were extremely limited. Most studies suggest that the aggregate reductions from carbon pricing were generally between 0 percent and 2 percent per year.

Two other studies not included in the meta-review reinforce the findings on the limited impact of carbon pricing. Both analyze taxes and trading jointly. One panel study of 149 countries across two decades finds that each euro increase in carbon prices was associated with a reduction in the *growth rate* of emissions by 0.3 percentage points.[25] In other words, this study finds that an increase in carbon prices, far from reducing emissions, only slowed the rate at which they increased. The other study finds that carbon pricing in thirty-nine developed countries (all in the OECD) resulted on average in a 1.5 percent reduction in economy-wide CO_2 emissions growth relative to a counterfactual. Put in other terms, the authors note that reaching the emissions reduction goals set forth in the Paris Agreement would require a global carbon price covering nearly 100 percent of emissions at more than $110 per ton.[26]

Conclusions for even some of the more successful programs carry some important caveats. The carbon tax in British Columbia managed to achieve reductions between 5 percent and 15 percent, relative to a counterfactual reference level. But the authors admit that this result did not account for the probable leakage to nearby jurisdictions.[27] Indeed, a more recent study finds the opposite: "Tax-induced aggregate emission reductions do not appear to be detectable."[28]

Third, taxes tend to do better than emissions ETSs. Although there is no causal explanation for this, one factor is likely to be the jurisdiction in which they function. Data from the World Bank show that 73 percent of carbon taxes are imposed at the national level. By contrast, only 40 percent of ETS occur at the national (or in the case of the European Union, supranational) level.[29] The rest are either subnational trading schemes, like those in California and Tokyo, or linked to international markets like the Paris Agreement Crediting Mechanism. Subnational schemes are more likely to have leakage problems, since there may be neighboring jurisdictions without a carbon price. In addition, firms prefer carbon trading over carbon taxes, since they see it either as a less costly form of regulation[30] or as one that they can more easily shape to their advantage.[31]

Fourth, and finally, the studies indicate that the largest and oldest emissions trading scheme, the EU-ETS, has had an extremely limited effect on emissions reductions. Even after years of learning and reform, most studies indicate that reductions are between 0 percent and 1.5 percent per year.[32] Admittedly, this result includes the early pilot phase, when the European Union was still figuring out how to design and implement the program.

After my study was published, there have been additional ex-post evaluations that also examine how carbon pricing affects emissions. The most significant is another meta-review that, like mine, analyzes only ex-post evaluations of carbon pricing schemes.[33] Niklas Döbbeling-Hildebrandt and his colleagues are considerably more sanguine about the effects of carbon pricing: they find that, across eighty evaluations of twenty-one different schemes, carbon pricing policies reduced emissions between 5 percent and 21 percent compared to a scenario with no carbon pricing. But their analysis raises a number of questions.

The authors' methodological approach is quite different.[34] Instead of using a standard regression model, they estimate the average effects of each carbon pricing policy using a method of probabilistic analysis called Bayesian model averaging. This approach is used when researchers do not have clear priors about what causes the outcome of interest—in this case, emissions reductions. Thus, the authors *predict* emissions reductions of a given ETS or tax by averaging the estimates of different models. However, there is already a large body of literature on how carbon pricing *actually* causes emissions reductions (or fails to do so). The authors' models contain different configurations of variables that they think may contribute to reductions, including the level of the carbon price, the policy design, and sectoral coverage, among other variables. The advantage of their approach is that they can then derive comparable averages across programs and do not have to preselect a statistical model. But it is unclear if such a comparison can be meaningful, given the huge variation in geography and policy design and the existence and stringency of other climate policies.[35]

The authors' methodological choice drives findings that contravene some of the empirical record to date. For example, their study finds that the Regional Greenhouse Gas Initiative (RGGI) has been a smashing success, reducing emissions by an average 21.05 percent compared to a scenario without this carbon pricing policy. However, leakage has been a perennial problem for RGGI: power generation has simply been shifted to states that are not participating in the program.[36]

Similarly, California's cap-and-trade scheme is estimated to have reduced emissions by 18.9 percent. But it too has experienced leakage, an extreme oversupply of credits,[37] and huge problems with offsets.[38] A number of researchers have pointed out that emissions reductions in California are primarily the result of a suite of *other* policies that the state has implemented.[39] Interestingly, when studies at risk of bias are removed from the model, reductions for the California ETS cannot be estimated. This indicates that robust data on its performance are simply not available.

Döbbeling-Hildebrandt and his colleagues report large reductions from China's pilot ETS, launched in 2013. These data also seem problematic, for two reasons. First, the Chinese pilot program was voluntary, suggesting that there may have been selection effects—jurisdictions that were planning to reduce emissions anyway were more likely to participate. Second, there have been ongoing problems with data reliability in the Chinese ETS.[40] If their average calculations are based on studies using inflated data, then they too will overestimate reductions.

Importantly, neither my meta-review nor the more recent one offers a causal model of what explains the level of emissions reductions (though admittedly, that is not the point of the exercise). But to have generalizable findings about the effects of carbon pricing, we need a clear theory of *how* carbon pricing works to reduce emissions—beyond a rote explanation that price increases curb pollution. In the following section, I explain how, in reality, carbon pricing favors fossil asset owners and therefore is unlikely to significantly reduce emissions, much less create new green asset owners.

Why Fossil Asset Owners Support Carbon Pricing

Managing tons aims to take the politics *out* of climate policy—by focusing on measurement, accounting, and trading—rather than confront the more contentious conflicts of asset revaluation. However, carbon pricing has become a controversial policy. Put bluntly, many voters don't like it, and political fights over carbon pricing have become quite common.[41]

Though the public at large often objects to carbon pricing, fossil asset owners tend to support it, since they perceive it as less costly than command-and-control policies.[42] Moreover, political backlash provides fossil asset owners with leverage to push for weaker carbon pricing policies or advantageous provisions. If governments must regulate emissions, carbon pricing is a preferred instrument of many fossil asset owners.

Fierce political fights over carbon pricing have emerged in several advanced economies. Australia's history of carbon pricing has been tumultuous; the policy has shifted with every change in leadership.[43] It now has the dubious honor of being the first developed country to repeal a carbon price. In the United States, a 2009 effort to implement a federal cap-and-trade program failed dramatically, despite promises to provide virtually all allowances for free.[44]

Subnational efforts in the United States have been checkered. Fourteen states have some form of carbon pricing in place.[45] However, passing legislation has been challenging in some states. The state of Washington finally approved the Climate Commitment Act in 2021, after many earlier failures,[46] and Oregon required two tries before approving "cap-and-invest" legislation. One of the attempts failed because Republican legislators fled the state to prevent a vote.[47] As I discuss in the following section, the story of California's ETS is one of lurching from problem to problem, with persistent difficulties around overallocation and low prices.

Other high-emitting countries have implemented carbon pricing but sub-sequently pulled back from price increases or slowed the rollout of the policy in some way. In 2016, Canadian Prime Minister Justin Trudeau implemented carbon pricing as part of the Pan-Canadian Framework on Clean Growth and Climate Change, hailing it as "the most cost-effective policy to cut emis-sions."[48] Since then, there have been multiple political battles over whether and how to implement carbon pricing in the provinces. Several provinces challenged the constitutionality of a carbon price; the case eventually went to the Supreme Court, which ruled in favor of the federal government. The law specifies that the price will increase by CAD$10 per ton each year, topping out at CAD$170 per ton in 2030—which, barring changes in other countries, would be the highest price in the world. In November 2023, Prime Minister Trudeau called for a temporary pause on applying the carbon tax to home heating oil, prompting a political pile-on from five provinces to expand the exemption to all home heating fuel sources.[49] The Liberal Party has since abandoned the consumer carbon tax altogether.

Political controversies around carbon pricing are not limited to these three nations. The riots by the Yellow Vests (*gilet jaunes*) in France were a response to an increase in fuel taxes (coupled with tax cuts for the rich) that was part of a broader strategy to reduce GHG emissions. The South African carbon tax passed after years of controversy in part because it offers generous tax-free emission allowances, ranging from 60 percent to 95 percent.[50] In 2022, South Africa extended these tax-free allowances.[51]

These political challenges appear to be at odds with the general narrative of managing tons, which, by emphasizing numbers and emissions accounting, tends to assuage opposition rather than provoke it. But in the case of carbon pricing, the most vocal opposition tends to come from voters and, as a result, the politicians who cater to them. Emitters tend *not* to be the biggest oppo-nents because, for them, carbon pricing is vastly preferred over other forms of regulation.[52]

In this view, carbon pricing as a policy of managing tons is triply problem-atic. Its effects on emissions are limited, and in many jurisdictions it is a tough sell politically: the costs of carbon pricing are not only up front but also very visible, often much more so than command-and-control regulations. In turn, these political challenges create leverage for fossil asset owners to pressure governments to weaken rules.

Most Likely to Succeed: The Mixed Performance of the European Union and California Cap-and-Trade Schemes

To put some of these critiques in context, I will examine two of the most likely cases for successful carbon trading: the EU and California ETSs. Both are highly developed economies with considerable governing capacity—for the European Union, at both the national and supranational (EU) levels. Second, both are regulated by centralized bodies that have been delegated extensive authority to ensure the proper functioning of the market. In California, the California Air Resources Board is responsible for all aspects of the market, including setting caps. In the European Union, the European Commission has control over setting caps for individual member states.[53] And in both jurisdictions, carbon pricing has historically not been as polarized an issue as in other advanced economies, like Australia or Canada. Of course, climate politics are extremely polarized in the United States at the federal level, but California has historically been a leader in environmental issues.[54] In short, if emissions trading succeeds, it should be in California and the European Union. But in both cases, the story is much more complex.

EU-ETS

The story of the EU-ETS is one of learning and reform, of identifying problems and responding to them, followed by gradual increases in program scope and ambition. While the largest carbon market is currently in a stable position with strong prices (€66 per ton in October 2024), its record on emissions reductions is mixed at best.

Overallocation has been *the* persistent challenge for the EU-ETS, often exacerbated by the use of international offsets from the Clean Development Mechanism. Its phased structure has allowed for important midcourse corrections that aim to control the supply of allowances to bolster prices.

Ultimately, the European Union took an unprecedented step to manage the ETS: it created a carbon central bank. The Market Stability Reserve (MSR) regulates the circulation of allowances, just as a central bank regulates the circulation of money. It appears to be making a significant dent in the twin problems of overallocation and price volatility. But getting to this stage has taken time, a lot of technical capacity, and a willingness to cede significant authority to a new regulatory body.

Created in 2003, the EU-ETS was up and running by 2005.[55] The pilot phase (2005–2007) covered only power generation and energy-intensive industries. Countries were allowed to set their own caps, which were then approved by the European Commission. Unsurprisingly, each nation was extremely generous in determining its cap, and the aggregate result was a huge overallocation and concomitant glut in allowances. Allowances were distributed for free. Prices fluctuated considerably—between zero and almost €30 per ton.[56] Essentially, countries spent this first trial period trying to figure out how to create a functioning market.

Several adjustments were made to the EU-ETS in each subsequent phase, responding to problems in the previous ones. With information in hand from the pilot phase about country-level emissions and the problem of overallocation, the European Commission took over the approval of allowances.[57] Moreover, by disallowing the use of Phase 1 allowances in Phase 2, the Commission effectively ended the overallocation problem—at least for the time being.[58]

The second phase of the EU-ETS lowered the overall cap and expanded it to include the production of nitrous oxide and aviation emissions (within the European Union only). However, most countries continued to receive free allocations (based on sectoral benchmarks) roughly equal to their emissions.[59] Banking was still not permitted, but the liberal use of extremely cheap offsets from the Kyoto Protocol's Clean Development Mechanism (see chapter 4) contributed to a glut in allowances, driving down prices.[60]

In 2012, as Phase 2 was ending, prices hovered around €5 per ton. Even more problematically, these allowances could be carried over to Phase 3.[61] By 2013, there was surplus of more than two billion allowances.[62] Prices plummeted to €2 to €3 per ton.[63] To respond to these problems, the European Commission implemented two additional reforms. For the first time, most allowances (57 percent) were auctioned rather than distributed for free.[64] In addition, the Commission agreed to reduce the overall cap by 1.74 percent per year, relative to the average Phase 2 cap, and to limit the use of offsets.[65] These corrections were undercut, however, by the ability to bank credits and by the fact that other climate policies continued to reduce the demand for allowances.

The solution to low prices and the glut of allowances was clear: remove excess allowances from circulation. In the short term, the European Commission agreed to slow down the number of allowances in circulation through "backloading"—essentially delaying the auctioning and distribution of some allowances until a later date. The backloading directive postponed the auctioning of 900 million allowances from 2014–2016 to 2019–2020.[66] Although backloading was a useful short-term fix for the immediate oversupply, it did not solve the underlying structural problem of too many allowances.

Enter the Market Stability Reserve (MSR). Created in 2014, the MSR began operation in 2019 to regulate the number of allowances in circulation.[67] The MSR regulates allowances in circulation through two mechanisms. It can withhold allowances through backloading. But more importantly, it can also cancel allowances altogether—essentially taking carbon currency out of circulation. These processes are governed by a formula: when certain prices or allowance numbers are reached, allowances are removed. When the total number of allowances in circulation exceeds 833 million, the MSR is triggered and 24 percent of allowances are withdrawn from future auctions and placed into the reserve for a year. Conversely, if allowances in circulation fall below 400 million, 100 million are released to be auctioned.[68]

The operation of the MSR corresponded to an increase in prices, though there have been debates about whether it might also destabilize the market through increased price volatility or speculation.[69] The technical discussions of the MSR's performance are growing, but at this point, they are largely in the realm of economic models. Though we cannot definitively point to the MSR as the primary cause, it is undeniable that prices have grown at a steady clip since 2019: as of February 2025, allowances are trading at around €70 per ton, down from €100 per ton in early 2023.[70] The role and reform of the MSR will undoubtedly figure prominently in the future performance of the EU-ETS; here the key point to emphasize is that the EU has delegated an incredible amount of authority to make this market function.

California

California is the other most likely successful case, and again, there is mixed evidence—despite ample regulatory capacity and political will. The primary political strategy has been to delegate the real work of policymaking and implementation to the California Air Resources Board. However, effective oversight of this body has been challenging, given the high degree of legislative consensus required. As a result, the California ETS has lurched from one problem to the next, without generating the steady incremental improvements that the EU-ETS was able to achieve.

A left-leaning state that has traditionally been a leader in US environmental policy, California passed AB 32, the Global Warming Solutions Act, in 2006. The goal was to cut emissions to 1990 levels by 2020 (now updated to net zero by 2045[71]). The bill was purposefully brief: it set goals without deciding on specific policies, and CARB was charged with developing and implementing the requisite regulations.[72]

The decision to put CARB in charge was strategic, intended to insulate the legislature from opposition while drafting the law and provide CARB with the necessary authority and autonomy to draft policies and quiet any interest group objections.[73] In theory, delegation to CARB seems like a good idea, but in practice it has created problems and uncertainty. California has struggled with extensive leakage, low prices, and likely inflation of offsets' contributions to reductions.

LEAKAGE

Trading began in 2013 and was immediately plagued by leakage. The lack of equivalent pricing policies in neighboring states created incentives to shift emissions-intensive activities out of state. Specifically, electric utilities engaged in "resource shuffling," using cheaper coal-fired power from Arizona and Nevada, despite a supposed ban on the practice in the regulations.[74] CARB maintains that there is limited evidence of resource shuffling, and that the carbon intensity of imported electricity has declined over the course of the ETS.[75]

Everyone else, however, appears more skeptical. The 2021 report of the Independent Emissions Market Advisory Committee suggested that CARB implement several measures to reduce leakage, indicating that this problem persisted.[76] California's Legislative Analyst's Office (LAO), a nonpartisan oversight body, noted that retrospective empirical data on resource shuffling were scant; as such, it would be difficult for CARB to make claims about abating resource shuffling.[77] And a recent study examining emissions from California and surrounding states estimates that the leakage rate in California is as high as 70 percent: reductions in natural gas emissions in-state corresponded to increases in coal-powered electricity in the western United States.[78] Attempting to control leakage by preventing resource shuffling is essentially enacting a carbon border adjustment tax, but utilities' political opposition has prevented the California State Legislature and CARB from successfully addressing this problem.[79]

OVERALLOCATION

Overallocation is the thorniest problem in the California ETS, resulting in persistently low prices. Firms participating in the market have logically taken advantage of the oversupply by purchasing cheap allowances and banking them for future use. Banking is a typical design measure to increase liquidity, but it has had perverse effects. One study estimates that the problem of overallocation is dire: banked allowances could permit firms to emit *in excess* of the state's cap while still complying with the market rules until 2030.[80] The Legislative Analyst's Office put the issue more bluntly: "The program is not stringent enough to drive the additional emission reductions needed because there will be more than enough allowances available for covered entities to continue to emit at levels exceeding the 2030 target."[81]

Excess allowances have, of course, affected prices. Between 2013 and mid-2021, prices remained below $20 per ton. Following the passage of AB 398, which extended the market until 2030, prices climbed modestly to $25 to $32 per ton until 2023, when they finally broke $35 per ton for the first time.[82] For reference, the High-Level Commission on Carbon Prices estimated that carbon prices would need to be between $40 and $80 per ton in 2020 and between $50 and $100 per ton in 2030 to limit warming to 1.5 degrees Celsius.[83]

OFFSETS

California currently allows regulated entities to use six different types of offset projects to satisfy a small proportion of their reductions. Until 2025, offsets could be used for up to 4 percent of total compliance obligations; beginning in 2026, this proportion climbs to 6 percent, until 2030.[84] Given the size of the overall market, offsets could thus make up 56 percent of the effect of cap-and-trade on emissions.[85]

Chapter 4 explains why offsets are a terrible idea. The main problem is the likelihood of overestimating the reductions of a specific project owing to both the difficulty of developing baselines and the political incentives to inflate them. California's mining protocols are a prime example. The mining cultivation offset protocol uses a standardized approach to facilitate estimates of additionality. However, given the wide variation in the size and methane concentrations of different mines, this approach greatly increases the risk of overcrediting (i.e., non-additional projects).[86] In the case of California, it may have even contributed to a decision by the federal government not to require methane capture for mining operations conducted on federal lands—effectively weakening national-level climate policy.[87]

Forestry offsets are equally problematic.[88] One study finds that forested lands with active offset projects had no less disturbance (in the form of deforestation or land conversion) than areas without projects.[89] This means that the offset credits literally have no beneficial climate effects. Another finds that risks to forestry permanence (mainly from wildfires) make "buffer pools"—the additional credits that buyers must purchase to guard against reversals in carbon sequestration—insufficient. Over the first decade of the program, wildfires burned at least 95 percent of the buffer pools. Thus, the authors note, "even if we make the implausible assumption that no additional wildfires will impact forest offsets projects in California's program, we nevertheless forecast that carbon reversals from historical fires will nearly drain and likely deplete the wildfire component of the buffer pool."[90] Quite simply, forestry offsets are a false solution.

IS CARB A ROGUE AGENT?

Despite multiple rounds of amendments to the rules, CARB has yet to solve these large problems. It has failed to explicitly acknowledge, let alone address, the dire oversupply problem.

Two independent review bodies, the Legislative Analyst's Office (LAO) and the Independent Emissions Market Advisory Committee, have made repeated calls for CARB to clarify how it will address overallocation. CARB has presented several plans, but the LAO has concluded that they are insufficient, and specifically that "cap-and-trade is not currently positioned to ensure the state meets it statutory 2030 GHG goal, much less CARB's more ambitious target [of 48 percent reduction below 1990 levels by 2030]."[91] Limiting banking or removing allowances from circulation are obvious solutions. California's ETS does have an "allowance price containment reserve," but this is intended to provide a safety valve for high prices rather than protect against low ones.

CARB has been delegated authority by the California State Legislature to make and enforce ETS rules. Insulating it from political pressure, the theory goes, will facilitate efficient and effective rulemaking and implementation. And yet clearly this has not come to pass. Why?

Theories of delegation assume that agents will carry out the wishes of the principals and that when they fail to do so the principals will get them to fall in line by exercising oversight, demanding accountability, or restricting or revoking agents' authority.[92] But these measures require political consensus among the principals. This is where things get tricky.

Proposition 26 in California states that all laws that increase taxes must be approved by a supermajority—that is, two-thirds of the legislature.[93] Since cap-and-trade effectively raises costs for regulated entities, it is considered a tax and therefore requires a supermajority. This supermajority was required to pass AB 398, which extends the ETS to 2030. But such moments are relatively rare, and they shift with elections, changes in political coalitions, and other generally unpredictable political considerations.

To improve oversight, AB 398 created the Independent Emissions Market Advisory Committee (IEMAC), which provides an annual analysis of the performance of the ETS. In its first report in 2018, the IEMAC immediately identified oversupply as a potential problem and has made recommendations annually about how to address it.[94] The repetition of these suggestions in subsequent annual reports indicates that CARB has largely ignored the IEMAC's input.[95]

The California ETS demonstrates the clash between market theory and political realities. Delegating authority to CARB helps insulate the legislature from political pressures to water down the ETS. But proper functioning of the market assumes that CARB will faithfully execute the legislature's instructions. When it fails to do this—especially regarding matters that can be construed as taxation—extraordinary political consensus is required for the principals to reassert control. Thus, a political analysis indicates that the institutional design of the California ETS will face repeated political challenges—and therefore is likely to face market volatility as well. Successful resolution, which would take the form of creating a central carbon bank, depends on the unpredictable nature of politics.

And Yet . . . Carbon Markets Continue to Expand

Despite the myriad implementation challenges, extensive regulatory requirements and limited impact on emissions, carbon pricing continues to expand, both domestically and internationally. And markets are increasingly linking together, creating another layer of political and policy challenges.[96]

As of late 2024, there were seventy-four carbon pricing instruments operating at the national or subnational level, covering roughly one-quarter of global emissions.[97] The majority are in the developed world, though some middle-income countries are now pursuing carbon pricing policies. New ETSs have recently been created or are being considered in Indonesia, Malaysia, Vietnam, and Japan, and the scope of existing ETSs has expanded in Mexico and the European Union.[98]

Article 6 of the Paris Agreement creates two new international carbon markets. Article 6.2 creates a framework for countries to enter into bilateral agreements to trade carbon allowances. Referred to as "internationally traded mitigation outcomes," these were a source of ongoing contention following the Paris Agreement. As of February 2025, forty-three bilateral agreements have been signed to begin the process of implementing Article 6.2 projects.[99] Switzerland has been particularly active in developing projects in Ghana, Senegal, Morocco, and Thailand and has signed bilateral agreements to the transfer of mitigation outcomes once the projects are operational.[100] As of November 2024, nearly a decade after the Paris Agreement was signed, only one project (between Switzerland and Thailand) has successfully transacted credits under Article 6.2.[101]

Article 6.4 creates a new offset market (see chapter 4). Contentious and highly technical debates delayed finalizing the rules for both mechanisms until 2021. NGOs and other advocates are still extremely concerned about the potential for creative accounting and bogus offsets in Article 6.4.[102] Although the CDM has been repeatedly and roundly criticized over concerns about additionality, the new offset market will grandfather in most CDM credits (all those generated after 2013).

Linkage

Another dimension of the growth of international carbon markets is linkage between subnational, domestic, and international markets.[103] Scholars and policymakers have championed the idea of linking carbon markets since the mid-2010s.[104] According to economic theory, linking markets should promote market liquidity and trading and lower overall costs.[105] Even better, these markets could provide an end run around long and fractious diplomatic negotiations while, ideally, providing convergence on a global or almost-global carbon price.

But the reality, of course, is much more complicated, in terms of both policy and politics.[106] For linkage to be effective, individual markets must be robust *before* they link—a condition that remains unfulfilled in most markets. If markets are not robust, linkage may propagate design problems across systems.[107] For instance, without coordination of levels of ambition, linking markets can create a race to the bottom. If one market is oversupplied (and allowances are therefore cheaper), linking may simply shift the purchase of allowances to the cheaper market, without reducing overall emissions.[108]

Political issues with linkage are even thornier. Linkage represents a loss of regulatory autonomy in the same way that the liberalization of any market does. Linkage may require one country to buy large amounts of allowances from another—in effect, a large transfer of wealth. Eventually, such requirements can become politically problematic.[109]

Governments can also choose to de-link markets, as the province of Ontario did. Ontario's cap-and-trade scheme was briefly linked to the California and Quebec markets, which joined together in 2014. Allowances were jointly auctioned and accepted in both markets. Ontario joined in January 2018 and then de-linked from the market six months later, under the leadership of Premier Doug Ford, who campaigned vociferously against carbon pricing. The withdrawal was expensive, costing the province CAD$3 billion, according to the Financial Accountability Office of Ontario.[110]

The European Union's Carbon Border Adjustment Mechanism

Perhaps the biggest development in the current carbon pricing policy landscape is the European Union's Carbon Border Adjustment Mechanism (CBAM). Designed to prevent carbon leakage in carbon-intensive sectors—cement, iron and steel, aluminum, fertilizer, electricity, hydrogen—the CBAM imposes a fee on the embedded carbon of certain imports that originate in jurisdictions without a carbon price (or those with a lower price). It aims to reduce unfair competition from markets with less ambitious climate policies.

Beginning in 2023, the CBAM requires an importer to obtain and surrender certificates equivalent to the carbon embedded in the imported goods. Consistent with all policies that manage tons, the calculation of embedded carbon is critical to the proper functioning of the CBAM. For the purposes of the tax, embedded carbon includes "Scope 1" and "Scope 2" emissions, which are generated through the direct combustion of fossil fuels (Scope 1) or indirectly through the purchase of electricity (Scope 2).[111]

Obtaining and verifying these data is a tedious affair. Ideally, embedded carbon will be measured at each individual production facility. Failing that, estimates will be based on sectoral benchmarks, which estimate the expected emissions intensity based on average performance by jurisdiction and available technologies.[112] However, even this is complicated. For example, a product using new steel made in a blast furnace (high carbon intensity) will have different average emissions from the same product produced using scrap steel in a much more efficient electric arc furnace.[113] Thus, the same product would have two different benchmarks, based on the production process. One possible workaround, using the average emissions intensity of the exporting country, is likely to run afoul of WTO rules unless carefully designed.[114]

As with all policies that manage tons, much will rest on the quality of the data. The transitional phase of the CBAM, which lasts until the end of 2025, intends to address the data issue by requiring importers to report the embedded emissions of their imports, without having to surrender certificates equivalent to these emissions. Much like the pilot phase of the EU-ETS, the data gathered during this transitional period will inform the measurement methodologies in subsequent periods, when certificates will be purchased and surrendered.

Even if these measurement issues can be addressed with minimal problems, the CBAM will require more regulatory infrastructure to oversee implementation. National registries will track importers and the CBAM certificates issued to them. At the supranational level, the EU Commission will create a centralized independent transaction log tracking the sale and repurchase of certificates.[115] And all producers must have their emissions data verified by a third party.

Finally, there is an additional complicating twist: the simultaneous phase-out of free allowances in the EU-ETS. Currently, EU firms in particularly carbon-intensive and trade-exposed sectors receive free allowances to reduce the possibility of leakage. The CBAM is designed to counteract the leakage problem from the other side of the equation by making *importers* comply with the costs of EU regulation. But simultaneous distribution of free allowances to EU exporters and imposition of the CBAM on EU importers will unfairly skew economic advantages to EU exporters, who will enjoy a double benefit. Thus, beginning in 2024, the free distribution of EU-ETS allowances will be phased out as CBAM requirements are phased in. In addition, the overall number of ETS allowances will be reduced by 4.3 percent annually from 2024 to 2027 and by 4.4 percent beginning in 2028.[116]

In other words, just as the EU-ETS has reached some stability, the goalposts are moving. Free allowances will be phased out, a new regulatory infrastructure will be created to administer the CBAM, and beginning in 2026, importers will have to surrender certificates to cover the embedded carbon of their goods. The hope is that the pilot phase will allow sufficient time for data gathering and learning to ensure that the numbers are actually correct—i.e., that importers can reliably calculate the carbon embedded in the goods coming into the European Union. Other jurisdictions are watching carefully, and a number of similar proposals are on the table in Canada, the United Kingdom, and the United States.[117]

A Flawed Policy That Is Here to Stay

Despite its many challenges, carbon pricing isn't going anywhere. New markets are being created, and existing ones are expanding and linking. This growth is occurring despite the fact that carbon pricing does relatively little to reduce emissions, often sparks opposition from voters and the public, and requires a high level of regulatory capacity. The European Union is arguably the most successful case of carbon pricing, but it has taken fifteen years, consistent improvements, and the creation of a carbon central bank to achieve its modest results.

The introduction of the CBAM creates a new phase of uncertainty and adjustment. California shows how delegation to insulate regulators from political pressure can backfire. Delegation requires vigilant and active principals to exercise oversight and ensure that agents are carrying out their marching orders, all of which requires political will. CARB has not addressed the overallocation problem, despite ongoing instructions to do so. Further action on allowances will require a supermajority from the legislature, which depends on many other political factors. As a result, California remains awash in allowances, and prices remain modest.

Danny Cullenward and David Victor call for a "right-sizing" of markets to recognize what they can and cannot do and avoid reliance on what they call "Potemkin markets" to achieve emissions reductions.[118] This point merits reiterating: collectively, we need to drastically scale back the use of markets, since they are likely to continue to deliver favorable treatment to fossil asset owners. As with any benefit, once in place, it is difficult to reverse. This is why surplus allowances, questionable offsets, and generous exemptions are the norm.

Put simply: carbon markets are slowing progress toward decarbonization. If carbon pricing must be used, levying taxes, not trading emissions, is the best path forward.[119] ETSs can be redesigned to resemble taxes, with limited use of banking, borrowing, and offsets. Price collars can specify the range of allowable prices to reduce volatility. These design choices, which make ETSs function more like taxes, are particularly useful in jurisdictions that may not have the legal or constitutional authority to levy new taxes.[120] Further linkages—including with the international markets created by Article 6 of the Paris Agreement—should not be pursued. As I outline in chapters 6 and 7, a better path forward is to redirect flows of capital in the global economy.

4

Carbon Offsets Are Fatally Flawed

THIS CHAPTER is for advocates, ESG officers and investors, and any organization or government using offsets as part of a net zero strategy. The message is simple, though it will undoubtedly raise many eyebrows: it's time to get rid of all nonpermanent offsets. Offsets are a good idea in theory, but they have failed to reduce emissions. Worse, they distract from the urgent task of decarbonization and provide ample opportunity for fossil asset owners to greenwash. Offsets are perhaps the textbook example of why managing tons is so problematic: despite serious questions about offsets' effectiveness, everyone wins as long as the numbers add up—everyone except the climate.

There has been increasing scrutiny of offsets in the last few years, particularly those in the so-called voluntary market. While journalists and the public may be surprised by the problems associated with offsets, for longtime observers of offset markets these are long-standing issues, many of which were baked in from the outset.

To understand why the problems with offsets are both unsurprising and not fixable, I first provide a primer on how they work. Many of us have a general understanding of offsets (which are also referred to as "carbon credits"), but as with all policies that manage tons, the details of measurement are complex—and essential for understanding offsets' fundamental flaws. I then turn to a history of this policy instrument from its origins as "debt for nature swaps" orchestrated by NGOs and "Joint Implementation" in the Framework Convention on Climate Change.

Next, I assess offsets' track record in reducing emissions, and show that they have underperformed from the start. The current menu of projects continues to face problems of scale, cost, and feasibility, as well as the fundamental problem of what should be considered an "additional" reduction in a time when we must be rapidly decarbonizing.

I then turn to the myriad political challenges posed by offsets and demonstrate why effective reform of this policy is not possible. Despite their questionable record on emissions reductions, most in the climate policy world are doubling down on carbon credits. COP29, held in Azerbaijan in 2024, finalized the rules of operation for the Paris Agreement Crediting Mechanism (PACM), which is the newest global offset market.[1] I detail how the growth of the unregulated voluntary market and the continued existence of a global offset market have created multiple implementation issues and entrenched constituencies championing offsets' expansion. This poisonous mix has created a monster—a ubiquitous but largely ineffective approach to climate change with lots of advocates and few prospects for meaningful reform.

The stark reality is that most types of carbon credits have far outlived their utility. They may have been a defensible approach when we were aiming to modestly reduce emissions in the early 1990s. But we are no longer in that world; now we must drastically reduce atmospheric concentrations of GHGs (not just emissions) and transform the ways in which we generate and use power. Bluntly put, offsets are incompatible with the Paris Agreement.[2]

The *only* offsets that should be permitted are those that permanently sequester GHGs, with no risk of reversibility. This means that "nature-based solutions"—such as planting trees or restoring forests—should no longer be considered legitimate carbon credits. We urgently need to protect and restore forests for a variety of important environmental benefits, but we need not do so through the structure of an offset project. As I detail in the final section, permanent removals are the path forward for carbon credits, but even those come with their own challenges.

A Primer on Offsets

Like other forms of carbon pricing, offsets create a new currency.[3] In theory, they commodify the absence of greenhouse gas emissions. But in practice, this commodity is actually the *hypothetical* absence of emissions. Project financing pays for "additional" reductions—those that would not have occurred without the project. Project developers estimate these additional reductions based on counterfactuals. While there are reams of complex methodologies that justify these estimations, ultimately they are fundamentally unverifiable claims.

These claims are then reviewed by a third-party validator at the project proposal stage and at the end of the project by a verifier. Third-party reviews may also be carried out during the project. In general, validators and verifiers are different organizations for a given project, to reduce potential conflicts of interest. However, both the project validator and verifier are paid by the project developer.

There are two key characteristics of offset projects: the timescale on which they operate, and the type of climate benefit they provide. An offset project can operate for a temporary or permanent period. Temporary offsets include renewable energy, agricultural, waste, and forestry projects, last a century or less, and comprise 99 percent of offsets sold. Permanent offsets sequester carbon on geological timescales through activities such as direct air and ocean capture, enhanced mineralization, and the use of bioenergy with carbon sequestration. These offsets are generated by emerging technologies that seek to directly remove carbon dioxide from either the atmosphere or the oceans and sequester it (almost) indefinitely.

Often, so-called permanent offsets are not, in fact, permanent. For example, forest preservation and expansion are generally considered to be permanent offsets. But as recent wildfires across North America and Europe have starkly illustrated, they are not. Such burns will become a greater problem because of the increasing likelihood of fires and droughts resulting from the intensifying effects of climate change. In 2023, forestry-related projects accounted for 35 percent of the voluntary market, down from 46 percent in 2022.[4] Thus, a large part of the market is at risk and cannot be regarded as permanent.

Offsets can offer two different types of climate benefits, depending on the project. They can either avoid emissions being added to the atmosphere or remove them altogether. A renewable power project will avoid the emissions that would be generated by gas- or coal-based power. An energy efficiency project also avoids emissions by reducing consumption. Carbon removals are most frequently generated through afforestation and reforestation or through direct air capture.

There are also two different markets for carbon credits: compliance and voluntary. Compliance markets are used by regulated entities to meet their emissions reductions requirements. At the international level, carbon credits were first created and traded through the Clean Development Mechanism (CDM), created by the Kyoto Protocol. Three-quarters of CDM projects are industrial gases and renewable energy.[5] The CDM's new incarnation is the Paris Agreement Crediting Mechanism, created by Article 6.4. The final rules were agreed to at COP29 in 2024, paving the way for this new offsetting market to become operational.[6] There are also many national and subnational markets for offsets, which I touch on later.

There is also a voluntary—or more accurately, an unregulated—market for carbon credits. (I use the terms "voluntary" and "unregulated" interchangeably.) These credits are governed by private standards, created by NGOs and other nonstate actors.[7] The unregulated market preceded the CDM by several years. As early as the 1980s, NGOs were engineering "debt-for-nature" swaps in which developed countries would cancel a portion of a developing country's debt in exchange for a promise to preserve some portion of its forests.[8] These eventually morphed into more sophisticated emissions reductions projects that have become the modern offset market.

The prominence of the unregulated market has risen and fallen over time, depending largely on the state of multilateral negotiations.[9] This market grew following the entry into force of the Kyoto Protocol and the implementation of the CDM. The CDM provided proof of concept, and NGOs bet that there would be a demand for offsets beyond the Kyoto Protocol. The voluntary market dipped after the (largely) failed COP15 in Copenhagen in 2009, but it has been on the march since the Paris Agreement was signed in 2015. One think tank that tracks the value of the voluntary carbon market estimates a 600 percent increase in value from 2018 to 2021. In 2021, the market was valued at around $2 billion. Since then, the market has contracted considerably, falling to $1.9 billion in 2022 and then to $723 million in 2023.[10]

Standard setters in the unregulated market are nonstate actors—generally nonprofit organizations—that develop the methodologies to govern offset practice (though in practice many of them are based on CDM methodologies). They also create rules about who can validate and verify projects and maintain registries of projects to enhance transparency.

Despite its growing value, the unregulated market is dominated by four main organizations: Verra (formerly VCS), the Gold Standard, ACR (formerly the American Carbon Registry), and the Climate Action Reserve. Verra is the clear leader in the market, responsible for 69 percent of all credits transacted in 2024, the last year for which comprehensive data are available.[11]

The growth in the unregulated market is driven by the increasing demand for offsets in all areas of climate policy. In 2015, countries signed the Carbon Offsetting and Reduction Scheme for International Aviation (CORSIA) agreement, which will cap aviation emissions at 2020 levels beginning in 2027. Since no carbon-free alternative to aviation exists, the majority of these reductions will come from offsets (with a much smaller amount derived from efficiency improvements and upgrades to aviation fleets). The International Civil Aviation Organization (ICAO) administers the agreement and estimates that it will require approximately 2.5 billion tons of CO_2 offsets between 2021 and

2035, representing an investment of at least $40 billion.[12] In 2020, governments agreed that certain carbon credits from the voluntary market could be used for purposes of complying with CORSIA—further embedding this unregulated market in the global climate policy landscape.[13]

As chapter 5 demonstrates, the net zero "wave" has also created a huge demand for offsets. Almost half of the largest publicly traded companies have now pledged to go net zero. The "net" in those pledges will come from carbon credits from the voluntary market. One recent study of almost five hundred major companies notes that "if these [firms'] plans are implemented, global demand for carbon credits is predicted to grow by several orders of magnitude."[14] Carbon offsetting isn't going anywhere.

Compliance markets are also still going strong, both internationally and domestically. The Paris Agreement has created new ways to manage tons through Article 6, as noted in chapter 3. And the PACM is poised to become a slightly reformed version of the much-criticized CDM, which will now include credits for removals.[15] This new global offset market is still in its early phases, but governments have already agreed to grandfather in credits from the CDM from 2013 forward.[16]

The domestic landscape of carbon offsetting is more heterogeneous and complex. A 2024 World Bank report on carbon pricing catalogs thirty-five carbon crediting mechanisms currently implemented globally from the national to the local level.[17] For example, Canada has a federal crediting mechanism created by its 2015 federal climate plan—the Pan-Canadian Framework on Clean Growth and Climate Change. In addition, five provinces have their own offsetting programs. Similarly, Japan has a national offsetting program, and the city of Tokyo has its own citywide cap-and-trade scheme that includes provisions for offsetting. As with the voluntary market, domestic and subnational offsetting is expanding, with another five programs in the implementation stage in Mexico, Chile, Canada, Indonesia, and India.[18]

In sum, carbon offsetting is everywhere. And its use—for both compliance and voluntary purposes—is only set to expand. I turn now to explain why this is such a huge problem for climate policy.

Implementation Problems

Like all forms of managing tons, offsets present an enormous accounting challenge. *They create a new asset whose value is based on the absence of emissions, estimated through counterfactuals.* As a result, the counterfactual, or "baseline," is critical. Most offsets are estimated against a baseline: How many emissions would have occurred without the financing for a given offset project? The difference between the "would have" world, and the "project" world is the amount of the carbon credit. Wonks will know this amount as "additionality"— the additional emissions reductions that occur as a result of financing for the project. Accurate estimation of additionality depends on the baseline estimation of emissions in the "would have" world. If this number is overestimated, then there's huge overcrediting. Put another way, offset buyers are getting credits for hypothetical reductions that are not likely to occur.

Additionality was and remains a huge challenge for the CDM (and for other domestic offset programs). Many critiques posit that the reductions generated by the CDM are hugely overestimated.[19] The California offset program has come under similar scrutiny.[20] The recurring discussions of additionality suggest that it is a challenge by design rather than a problem specific to one offset program.[21] A recent meta-analysis of over two thousand carbon offset projects, covering 20 percent of total credits issued over the life of offset markets, found that "less than 16% of the carbon credits issued to the investigated projects constitute real emission reduction."[22] As I explain later, given the design and political interests of those involved in offset programs, this should not be surprising.

Second, and in addition to the fundamental problem of baselines, there is a more subjective question of what types of projects should count as "additional" in a decarbonizing world. Should companies get credits for funding projects that switch fuels from coal to natural gas? Currently, this is an allowable project under the CDM rules.[23] What about "avoided deforestation"? These are projects demonstrating that areas would have been deforested or degraded without project funding. But this creates the perverse incentive to threaten to deforest areas to receive funding to "preserve" them. Indeed, during the creation of the CDM, NGOs advocated vigorously against the inclusion of forestry offsets for precisely this reason.

Arguably, in a world of increasingly stringent climate policy, activities like fuel switching, energy efficiency, or use of renewables should not be different from "business as usual"—the baseline estimate for offset projects. Many of these types of emissions-reducing activities are already becoming business as usual. Some offset providers in the voluntary market have adjusted accordingly, disallowing certain types of offset projects. But this change is far from universal. As a result, many offset programs continue to allow project types that arguably comprise activities that should be part of a new status quo in a climate-changing world.

Third, leakage is another basic problem. Offsets are generally implemented as discrete projects, meaning they have geographical boundaries. Thus, a project is designed to avoid deforestation in a specific tract of land, or an individual electricity plant switches from coal to natural gas. But it is again difficult, if not impossible, to ascertain whether the projects have simply caused carbon-emitting activities to be shifted beyond its borders. The resulting leakage means no net emissions reductions.

Fourth is the question of permanence. "Nature-based solutions" are all the rage in the voluntary market. This umbrella term includes the protection and restoration of ecosystems that provide useful mitigation and adaptation functions—ranging from reforestation to restoration of wetlands (to protect against flooding). Yet these are not permanent solutions. In the summer of 2023, forest fires in Canada were estimated to have released approximately two billion tons of CO_2—approximately triple the country's annual carbon footprint.[24] As the world faces an increasing number of climate-related extreme weather events, the permanence of forestry credits will be threatened.

In fairness, some offsetting methodologies plan for this possibility by requiring "buffers" in their projects. In essence, buffer credits are a form of insurance. They require the buyer to pay for more credits than it will actually claim, in the event that some carbon stocks are lost.

Fifth is the wooly accounting challenge of double counting. This problem gets deep into the weeds of Article 6 of the Paris Agreement, which prevents most double counting through its "corresponding adjustments" requirement.[25] Corresponding adjustments mean that the country selling an offset cannot claim those reductions; only the buyer can. Thus, the seller must adjust its reported emissions; it cannot deduct the amount of the offset from its overall emissions.

But the voluntary market is not subject to corresponding adjustments. In theory, then, if a firm in Spain purchases an offset generated in Chile, both the Spanish firm and the Chilean provider (or government) can claim the reduction. If both entities claim the reduction, then it is double-counted. Some voluntary offset standards are choosing to use corresponding adjustments, but this is not required. Offsets that do not use corresponding adjustments are directly undermining the goals of the Paris Agreement.[26]

Optimists suggest that this problem will be resolved, since eventually buyers will only want to purchase "higher-quality" credits—that is, those that have participated in the process of corresponding adjustments.[27] But there is no guarantee that this will be the case.

Worse still, double counting can occur just through the different ways in which actors keep their books. Double counting occurs if the same emission reduction is counted twice, either by sellers (double issuance) or by different buyers (double claiming). Double claiming occurs when a country's emission reduction is counted in its own inventory and transferred to another country without proper accounting, allowing both countries to use it toward their mitigation pledges. Double issuance happens when two different entities issue a unit for the same emission reduction. In both instances, a centralized registry must oversee international transfers of credits to avoid double counting.[28]

Sixth, problematic rules are reproduced throughout the landscape of public and private rules. Most offset standards in the unregulated market are based in part on CDM methodologies.[29] But as noted earlier, there are many concerns about whether these methodologies actually produce additional reductions beyond business as usual. And to the extent that the unregulated market is built on faulty rules, it will reproduce these problems of non-additionality.

Seventh, carbon credit prices are simply too low. Part of the logic behind offsetting is to generate investments in jurisdictions in need of funding for an energy transition. In mid-2023, the average price for most offset types was around $5 per ton (excluding permanent removals, which I discuss later).[30] Other carbon trading platforms, such as AirCarbon Exchange and CBL, listed offset prices well under a dollar in late 2024. More specifically, futures contracts for nature-based offsets were trading for US$0.32—thirty-two *cents*— per ton, far less than the cost of a cup of coffee. Even with the huge quantities of credits transacted, this is not the kind of money that will drive decarbonization. Moreover, this price pales in comparison to the economic damage caused by climate change—or the "social cost of carbon"—which is used to craft government regulations. Estimates of this cost vary widely by country; one recent study puts the global mean estimate at US$185 per ton.[31] Of course, there is no "magic number," and some studies suggest that the social cost of carbon could be as high as $1,000 per ton.[32] There is considerable debate about the "right" number.[33] (This huge variation again speaks to the implementation challenges of managing tons.) Debates aside, carbon credit prices are obscenely low.

And finally, there is a nontrivial problem of fraud. In October 2024, the US government charged Kenneth Newcombe, a former World Bank employee, Verra board member, and longtime advocate for carbon markets, with fraud related to carbon offset projects. Along with other employees at C-Quest Capital, Newcombe allegedly fabricated data about the emissions reductions achieved through twenty-seven cookstove projects in Zambia and Malawi.[34] These projects, listed on the Verra registry, have been suspended, and it is still unclear what will happen to the credits and whether the investors will be compensated.

The C-Quest case is not an anomaly. There have been other high-profile fraud cases in the voluntary market. South Pole, another project developer, summarily cut ties with an enormous forest conservation project in Zimbabwe after allegations of fraud.[35] The Zimbabwe project, which also used Verra off-set protocols, generated over thirty-six million credits over the lifetime of the project.[36] (It is worth noting that the financing for this fraudulent project often came through tax haven countries like the Cayman Islands; I discuss tax havens in further detail in chapter 6.) Two additional Verra projects in the Brazilian Amazon have been accused of laundering timber that was harvested illegally. As of late 2024, those projects have been suspended, and Verra is conducting an internal review.[37]

Optimists will counter that every policy has implementation problems, and that reform can address many of these issues. Indeed, there have been changes to the CDM as well as multiple efforts to improve the "integrity" of the voluntary market. It is telling that these reform efforts resurface every few years, indicating the persistence of many serious problems.

In 2008, the International Emissions Trading Association (IETA)—an industry association promoting emissions trading—created the International Carbon Reduction and Offset Alliance (ICROA)—yet *another* standard-setting organization to "enhance integrity in the voluntary carbon market."[38] Standards on the voluntary market can now signal their quality by saying that they adhere to ICROA best practices.

Yet ICROA was insufficient to allay concerns about quality. Fifteen years on, a *new* private initiative was created to achieve the same goal. The Integrity Council for the Voluntary Carbon Market (ICVCM) has created ten "core carbon principles" to ensure that projects meet the highest standards for additionality, sustainability, and governance. Released in 2023, the principles are accompanied by a set of decision tools to operationalize them. The ICVCM has begun approving both carbon credit programs and individual project methodologies aligned with the core principles.

But effective reform is not possible for the simple reason that the actors involved have no interest in any fundamental changes to the design and use of offsets as a policy tool. As Cullenward and Victor succinctly note in their discussion of offset markets, "there is no constituency for quality."[39] The absence of constituencies advocating for real reform is a consequence of managing tons. Since the challenge of emissions reductions has been reduced to more agreeable issues about the commodification of carbon emissions, there are no losers in the use of offsets.[40] And when everyone wins, there is little incentive for reform.

The political obstacles to reform fall into three categories: (almost) everyone wins using offsets; too many well-established and politically powerful actors want offsets to continue; and the lack of financial or reputational reward for "high-integrity" credits drives a race to the bottom.

Everyone Wants Big Numbers

The institutional landscape for carbon offsets includes many different actors, all of whom want the numbers on reductions to add up. Sellers want to maximize the number of credits they can produce. Buyers generally want cheap credits (though some have argued that this is changing). The middlemen—retailers, validators, and verifiers—want to make sure that the deal goes through.

And pretty much everyone values quantity over quality. For instance, project developers have an incentive to invest in projects that are "just barely infeasible"—that is, that will get credits for minimal investment.[41] This is why there is hardly any investment in permanent offsets like direct air capture (DAC), which has an immediate and permanent climate benefit. Currently, however, these offsets cost between $600 and $1,000 a ton.[42] If you run a firm that engages in greenwashing, why spend money on "boutique" carbon credits when you can buy "Walmart" credits much more cheaply?[43]

Developers outline a project's scope and implementation in the project design document, which includes the methodology used, estimates of reductions, plans for monitoring and evaluation, and an assessment of environmental and social impacts. They then secure funding and seek approval for the project to move forward—from either a government (in the case of a compliance market) or a private regulatory body (a voluntary market). At the international level, approval is granted by the Clean Development Mechanism Executive Board and now also by the Supervisory Body for Article 6.4 of the Paris Agreement.

In general, project developers aim to maximize the additionality of the project and therefore the number of credits they can sell. There are some built-in checks. A project must be validated by a third party at the outset and verified at the end of the project. The validator reviews the project plan, its methodology, and its estimates about additionality. Validators are accredited by the government or CDM Executive Board in the case of compliance markets, or by the private standard-setter in the case of the voluntary market. The validator is paid by the project developer, so it too has an incentive to approve projects from its client, for the maximal number of credits.

The verifier reviews the project at the end of its life to ensure that all the project activities were successfully carried out and the emissions reductions credits generated. It too is paid by the project developer, thus creating incentives to greenlight and approve projects with generous estimations of reductions. In most markets, the validator and verifier must be different entities. However, since these organizations often validate and/or verify each other's work, there is an incentive for verifiers to say yes to projects validated by organizations it may repeatedly encounter.[44]

These detailed review requirements are about *process*, however, rather than content. Thus, validators and verifiers make sure that the projects follow the rules, but not whether the rules themselves (i.e., the way that reductions are calculated) are actually robust. A "good" carbon offset project can therefore have bad methodologies. It can conform to procedures to promote quality without reducing emissions, or reducing much less than it claims.

Standard-setters in the voluntary market have an additional reason to want the numbers to add up: they need people to buy their credits. Elsewhere I have referred to the voluntary market as a form of private authority: when NGOs and other nonstate actors "make rules or set standards that other actors in world politics adopt."[45] Earning this authority requires negotiating a tricky balance. Attracting users is critical; without buyers, privately created offset standards are irrelevant. Thus, voluntary market rules must be sufficiently credible to attract users, but not so costly that they deter potential rule-takers. This is an ongoing balancing act: overly onerous rules will chase away the source of their authority—credit buyers in the offset market. But rules must be sufficiently credible to attract buyers; hence the repeated reform efforts to ensure integrity.

Entrenched Pro-Offset Interests

Carbon markets are now big business. The voluntary market is currently valued at almost $1 billion (it peaked at $2 billion in 2022), and the CDM market channeled $300 billion into climate and sustainable development projects between 2001 and 2018.[46] These big numbers create powerful political interests, including big corporations that want to buy credits, private regulators that want to supply them, and the validators and verifiers that monitor them. As a result, and despite the multiple and very public critiques of the unregulated market, it is growing *rapidly*—not only in value, as noted earlier, but also in its reach.

Credits from the unregulated market can now be used for compliance purposes in CORSIA, the aviation agreement. This did not happen by accident. Private standards like Verra and Gold Standard lobbied for this provision, as did IETA, the industry association that promotes emissions trading.[47] The acceptance of offsets from the voluntary market is a huge victory for private standard-setters—both in terms of legitimating their authority and generating demand for their products.

This trend is set to continue as the lines between the unregulated and compliance markets become increasingly blurry. A number of offset providers in the voluntary market are trying to extend their reach by contributing to the implementation of Article 6. The Gold Standard, which currently represents 10 percent of projects by value on the voluntary market,[48] has started a pilot program to help project developers with Article 6 governmental authorizations.[49] And the head of environmental markets at Gold Standard has put the matter more plainly: "Let's align the voluntary carbon market with Paris rather than play by our own rules."[50] Similarly, Ecosystem Marketplace, an NGO that provides information about the unregulated carbon market, has partnered with the US Department of State to provide "information and capacity building" to developing countries so that they can better understand how offsetting can help them meet their nationally determined contributions (NDCs).[51] True to the political playbook, once private rulemakers gain authority, they engage in multiple strategies to expand it.[52]

Oversupply

Third, just as fossil assets must be retired to decarbonize the economy, so too must carbon credits. Once a credit is retired (also referred to as "canceled" or "deregistered"), it is no longer included on the buyer's balance sheet as a reduction. Retirement also means that the credits cannot be resold to another entity. Resales look good on paper. They increase the value and activity of an offset market, but they do not represent more reduction (or avoidance) activities—simply the transfer of ownership. Rather than shuffling credits to the next owner, who can then claim the same reduction activities, cancellation leads to more investment in new reduction or removal activities.

In the voluntary market, issuances vastly outstrip retirements. In 2023, roughly four times as many credits were issued than retired, resulting in a huge surplus of credits.[53] There is no requirement in the voluntary market that credits be canceled, further exacerbating the surplus problem—and driving down prices.

In a (very slight) nod to concerns about oversupply and additionality, countries subsequently decided that they would cut down on concerns about non-additional projects in the Paris Agreement Crediting Mechanism by only allowing projects created after 2013.[54] This would weed out the projects most likely to be non-additional. While older projects are more likely to be non-additional, the problem is not unique to these credits.[55] The problem of non-additionality or overcrediting is rife throughout offset markets.

Article 6 also requires transparency about the vintage of CDM projects. The logic here is that buyers will have more information about which projects are likely to be of higher quality and will choose to purchase those. But if buyers are simply looking to balance the books, they may not be deterred from buying low-quality credits.

Both the extremely modest requirements for retiring credits under the new offset mechanism and the grandfathering of CDM credits indicate that the profound problems with offsets are being baked into the new cake. Despite the many problems with the implementation of the CDM, only the most minimal reforms have been put in place, and they will not solve the many fundamental problems I have detailed here.

Can Offsets Be Salvaged?

It is clear from this litany of implementation challenges, coupled with the political forces that further entrench the status quo, that global climate policy has a serious offsets problem. I am not the first to argue that the offset market cannot be fixed.[56] Any serious efforts to achieve the goals of the Paris Agreement *must include weaning ourselves from all nonpermanent offsets as quickly as possible*. Instead of counting tons, actors can disclose the investments that they have made to decarbonize.

Limiting credits to permanent removals requires phasing out avoided emissions projects as quickly as possible. Afforestation and reforestation projects should end. These projects are problematic for many reasons: they have repeatedly been shown to generate junk credits;[57] their prices are infeasibly low; and they dominate the voluntary market. These credits need to be retired as quickly as possible from both compliance and voluntary markets. No new avoided emissions projects should be permitted, and they should be withdrawn as permissible methodologies from both compliance and voluntary markets.

Second, the voluntary market must be regulated. Despite now being used for compliance purposes in the aviation agreement (CORSIA), the voluntary market is entirely self-regulated. And the cracks are showing. Unchecked growth of the voluntary market would be harmful on many levels, producing more bad credits with questionable additionality and incredibly cheap prices. A 2023 report by Bloomberg NEF notes, "If all offset types are allowed, the [voluntary] market will be oversupplied, and prices will average just $18/ton out to 2050. Conversely, a removal-only market would increase this average to a massive $127/ton."[58] The report adds that without significant reform, "prices would be *criminally low* in the voluntary market: Supply would be almost four times greater than demand in 2030."[59]

In addition to vastly restricting the types of allowable projects, a regulated voluntary market must deal with the potential for double counting. One approach is to require that offsets from the voluntary market be subject to the same "corresponding adjustment" requirements used in the Paris Agreement. These adjustments prevent the double counting of credits, ensuring that only the buyer can claim reductions to its balance sheet. While this is one possible solution, it has measurement challenges of its own related to the timing of reductions versus the accounting methods used.[60]

Another possible approach is to shift the voluntary market toward a "contribution" model. Instead of purchasing credits to justify (that is, offset) emitting activities, entities would invest in projects and simply report how the project contributes to reducing GHG emissions. As some of its proponents note:

> This change in terminology may seem small, but it represents a fundamentally different approach. For one thing, not allowing companies to subtract carbon credits from their direct emissions into a single net number, as offsetting does, refocuses priorities on direct emissions reductions. Companies would no longer be able to hide inaction behind offset purchases.[61]

Such an approach would also reduce incentives to invest in low-quality projects simply to maximize credits.

And finally, we should not be creating new kinds of carbon credits. Small start-ups are exploring the use of coastal and ocean offset projects to sequester carbon in biomass and soil in marine ecosystems. Some private standards have already approved methodologies to expand these types of projects.[62] These should not be permitted in either voluntary or compliance markets, as they will just entrench more pro-offset interests and add to the need for future reform.

Of course, given the forces that militate toward the status quo and the fact that offsets are now a pervasive part of many climate portfolios, generating political incentives to implement these reforms is the challenge. Some governments are exploring options for regulating the voluntary market. For instance, the US Commodity Futures Trading Commission had been exploring ways to regulate both derivatives and spot markets for offsets (though this is likely to change under the second Trump administration).[63] These measures can help further delegitimate the use of dubious offsets, which are already driving some firms to reduce their reliance on them.[64]

Scholars, investigative journalists, and NGOs are continuing to provide more evidence on the myriad problems with carbon offsets. This is an important additional push-factor to delegitimize offsets, particularly those in the unregulated market. A growing number of stories have appeared in the popular press about junk credits.[65] And while many NGOs are active supporters of offsets (and carbon markets more generally) as an important climate solution, a vocal minority opposes them. In October 2021, in advance of COP26, more than 170 NGOs wrote an open letter titled, quite simply, "Offsets Don't Stop Climate Change."[66] These types of efforts continue to create pressure to reexamine the role of offsets in implementing the Paris Agreement.

As with all climate problems, there is no magic bullet for generating solutions. But the first step is to recognize that offsets are a fundamentally flawed policy that allows actors and institutions to claim they are acting responsibly when they are not. We can no longer fudge the numbers. It's time to get rid of all nonpermanent offsets.

5

Net Zero

AN ELABORATE DISTRACTION

ACHIEVING NET ZERO emissions is now the guiding principle for global climate policy, and many fossil asset owners are making big promises. Saudi Arabia, home to some of the largest oil reserves in the world, has pledged to go net zero by 2060. The steel manufacturer ArcelorMittal has set 2050 as its net zero goal. Even Canadian oil sands producers—which extract some of the most carbon-intensive oil in the world—have a plan for going net zero.

Pledges sound nice, but a closer look raises cause for deep concern, grounded in the fundamental problems of managing tons. In this chapter, I show that the measurement and implementation challenges associated with net zero are profound, particularly for firms.

Existential politics implies that the problems with net zero cannot be solved. As long as fossil asset owners run the show, they will slow-walk decarbonization by engaging in practices that detract from the real problems of tackling supply.[1] Indeed, it is telling that only 15 percent of governments have committed their net zero pledge to law.[2] Even fewer are committed to ceasing fossil fuel exploration. Thus far, net zero is an elaborate distraction from the cold realities of asset revaluation.

If firms are really getting serious about net zero, they must accurately measure their emissions all along the supply chain—their so-called Scope 3 emissions. A true understanding of these carbon accounting challenges prompts the question: Does it even make sense for a firm to try to be "net zero"? At best, this is a herculean task; some would say it is simply not possible. As I explain later in this chapter, net zero accounting is rife with conceptual challenges that make "accurate" calculations of emissions difficult.

The problem with net zero goes deeper, however, than accounting issues. It lies in the very construction of the concept. "Net" means balancing emissions with an equal amount of CO_2 (or other GHG) removed from the atmosphere. Scientifically, there is a global net zero, where emissions are balanced out by removals. In theory, governments can coordinate to balance their collective emissions with removals, though even this undertaking is challenging. But at the level of individual organizations, I show that the logic of net zero breaks down.

Even at the national level, net zero should be carefully examined. The national inventory guidelines that form the basis of emissions reporting have not been formally updated in more than a decade. Moreover, these guidelines are based on production rather than consumption; many maintain that the latter is a more appropriate metric.

Currently, most of the "net" in net zero comes from offsets (offsets are now being referred to as "credits used to make offsetting claims"). As I documented in chapter 4, even without the problems of gaming measurement, the variety of approaches to offsetting emissions presents its own problems.

These measurement challenges are not flukes; they are baked into the cake of managing tons. Many "tons-related" problems, like assessing offset additionality or quantifying supply chain emissions, are effectively unsolvable. But this fact has not prevented fossil asset owners, governments, and eager NGOs from expending huge amounts of resources to try to address them.[3]

To fully understand the origins of some of these problems, I first examine the history of net zero. The concept comes in many flavors. These different names and definitions have created collective confusion about what net zero means, creating space for opportunistic interpretations.

Making a net zero pledge requires a deep understanding of carbon accounting. In this chapter, I first turn to the nuts and bolts of managing tons. Firms must try to calculate at least some of the Scope 3 emissions generated along their supply chains. And governments must wrestle with the accounting problems associated with embodied carbon and double counting. As with other forms of managing tons, the technical nature of net zero works in favor of the pledger, who has ample opportunities for strategic construction of pledges to make them appear more ambitious than they are in practice.

I then present data about who is doing what. There has been a proliferation of net zero pledges over the last four years, particularly by governments. These data affirm that net zero has become a norm in climate policy. To understand fully what this norm means, we must also examine how robust net zero practices are. I present findings from new research with colleagues that indicates reasons for concern. Unspecified reliance on offsets and a lack of accountability mechanisms—both of which are prevalent in pledgers' practices—can undermine the robustness of their net zero pledges.

Making sense of pledges is further complicated by the many rules governing net zero. There has been an explosion of voluntary net zero standards and guidelines; these are all ways to figure out how to "fix" the fundamental problems that invariably follow from managing tons. I describe two of the most prominent initiatives—the Science Based Targets initiative (SBTi) and the UN Race to Zero campaign. Then, to understand whether these pledges are mere exercises in greenwashing, I draw from literature on global governance to explain the inherent strengths and weaknesses of these private forms of authority. Ultimately, as with other forms of managing tons, the preoccupation with measurement and reporting distracts from the existential politics of climate change.[4]

The final section examines the political problems with net zero, showing why it looks mostly like an elaborate distraction. Weak standards are clearly a problem, compounded by the reliance on problematic offsets. But most importantly, many pledges—from both governments and companies—fail to address the elephant in the room: phasing out supply.

The Scientific Origins of Net Zero

To understand net zero's many challenges, we begin at the beginning. Before it became the North Star of multilateral climate policy, net zero was the stuff of ecologists, biologists, and other natural scientists. Net zero is fundamentally a scientific concept, applicable only at the global level.[5] Simply put, net zero means a world in balance, with all residual emissions (those that cannot be abated) "canceled out" by an equal amount of removals.[6] Any net zero pledge at the subglobal level cannot, by definition, be the same as scientific net zero. A subglobal pledge is a matter of assigning emissions to a firm or nation and then balancing them with offsets or removals. As I explain later, such a pledge, at bottom, is a subjective endeavor.

The "canceling out" comes from terrestrial carbon sinks and offsets. For this reason, the concept of offsetting has been part of net zero for decades. Early research examined fuel combustion offset by carbon sequestration,[7] or the quantification of emissions "saved" through the use of biomass-based energy.[8] The 2000 IPCC Special Report on Land Use, Land Use Change, and

Forestry repeatedly referenced "balancing," whereby carbon uptake is balanced by emissions from land use change.[9] Early on, as balancing became a theme in research,[10] "balancing" and "sinks" (especially forests) were coupled in discussions about carbon neutrality. Net zero is simply balancing where the balance is zero.

Net zero's transformation from scientific concept to policy principle helps explain why there is so much confusion about what net zero actually *is* and, importantly, how to achieve it.[11] First, the concept has always had multiple names. In the early and mid-2000s, "carbon neutrality" was the term of art: several countries, including Bhutan, Costa Rica, the Maldives, Morocco, and Papua New Guinea, adopted "carbon-neutral" targets in their national plans. The Fourth Assessment Report of the Intergovernmental Panel on Climate Change, published in 2007, refers to "zero net emissions," as do many articles around the same time period.[12] Indeed, the IPCC does not invoke the term "net zero" until its most recent assessment report.[13]

Second, net zero has different definitions. In the post-Paris era, it is often defined as emissions pathways that are consistent with 1.5 degrees Celsius of warming.[14] For example, the Science Based Targets initiative, one of the most prominent sets of voluntary corporate net zero guidelines, notes that net zero is achieved when firms reduce emissions and residual emissions consistent with eligible 1.5 degrees Celsius scenarios.[15] But a more rigorous approach defines it as "net-zero CO_2-equivalent emissions aggregated using the 100-year 'global warming potential' metric."[16] Global warming potential characterizes the relative warming effect of each greenhouse gas over one hundred years. CO_2 is the reference point, with a global warming potential of one; all other GHGs are measured in comparison. This definition of net zero is not associated with a particular temperature path.

Third, not all net zero claims are created equal. Some pledges include only CO_2, while others include a broader basket of GHGs. This is significant because GHGs vary in their life span in the atmosphere. For example, methane is a "short-lived" but extremely powerful pollutant that lasts in the atmosphere about twelve years on average. It is the second-most abundant GHG.[17] Forty percent of methane emissions are generated by agriculture. JBS, one of the largest multinational agricultural firms, has made a net zero pledge. Yet it does not report, much less explain, how it plans to reduce the emissions associated with livestock in that pledge.[18]

Fourth, the scope of the sources of emissions covered also varies. Company targets might cover only their "direct" emissions, such as those generated by the consumption of electricity, corporate vehicles, or purchased goods and services. These are referred to as Scope 1 and Scope 2 in corporate emissions accounting. Alternatively, they can also include some or all indirect emissions generated upstream in the supply chain. So, for example, a car manufacturer measuring its Scope 3 would include all emissions generated by, say, the steel used as inputs for the car (upstream, indirect) as well as those generated during the lifetime of each car (downstream, indirect). A bank's Scope 3 emissions would include emissions generated through firms supported by bank loans.

Finally, time frames are important too. While net zero is the midcentury goal, interim targets are important for several reasons. They help keep pledgers on track, ensuring that their long-term goals are consistent with their short-term actions and that they are not leaving reductions to the "last minute." Meeting interim targets also provides short-term climate and economic benefits. We know that the longer we postpone reducing emissions, the more expensive and difficult these reductions become.[19] Interim targets are a cost- and climate-control measure in addition to an accountability mechanism.

Net Zero Accounting

ExxonMobil, hardly an environmental leader, made headlines when it announced that its "operated assets" would be net zero by 2050.[20] How does an oil and gas company hit net zero in roughly twenty-five years? Its most recent plan includes measures such as reducing methane flaring, electrifying its drilling fleet, and offsetting emissions through carbon capture and storage.[21] But it also includes increasing energy supply. This is the fundamental flaw of the logic of net zero: the "net" effectively allows firms to postpone solving the thorny issue of asset revaluation.

Buying offsets or relying on carbon capture and storage allows fossil asset owners to continue generating profits (though these may be reduced due to investments in mitigation and offsets) and therefore maintaining their political power. Equally problematic, the many challenges of carbon measurement enable these fossil asset owners to "balance the books" on paper while continuing to emit.

As chapter 4 demonstrated, measuring emissions sounds straightforward, but in practice it is an incredibly complex task. A full understanding of the ambition and completeness of firms' net zero pledges requires a small detour into still more details of carbon measurement. The next section discusses the different challenges faced by organizations and states in accounting for their emissions. States measure overall emissions by sector, while firms measure their direct and indirect emissions. Both face significant implementation and political challenges.

National GHG Inventories

Countries have been reporting on their emissions since the Framework Convention was signed in 1992. The UNFCCC requires states to develop and report their national inventories of GHG emissions, including carbon sinks (Article 4.2). Correspondingly, the IPCC issued its first guidelines for national-level accounting and reporting in 1994.[22] Thus, countries have decades of experience in calculating and reporting GHG emissions—an obvious prerequisite for accurate net zero accounting. And yet multiple challenges for accurate reporting remain.

First, while the IPCC Guidelines provide a standardized approach for measurement and reporting, they cannot eliminate all the uncertainty associated with the process.[23] Second, emissions data are self-reported, and therefore concerns about misrepresentation abound. For example, a study comparing two sources of data for Chinese emissions—one national and one a compilation of provincial data—found a discrepancy of 1.4 gigatons of CO_2e in 2010—greater than the total emissions of Japan.[24]

Third, despite being the basis for measuring our collective progress on mitigation, the IPCC Guidelines for national emissions inventories have had limited updates. Initially drafted in 1994, the Guidelines were revised in 1996, updated in 2006, and "refined" by the IPCC in 2019. As the IPCC notes, the 2019 changes do not replace the 2006 Guidelines but rather provide updates, supplements, and elaborations as needed.[25] The IPCC has issued "good practice guidance" in some areas, such as wetlands and land use change, but not since the early 2000s. This has led to questions about whether the guidelines incorporate the latest scientific research and knowledge.[26]

And finally, there is a debate about whether national inventories should measure only those emissions produced within their boundaries or also include the emissions associated with imports.[27] The latter approach also accounts for international trade in "embodied carbon"—the emissions generated through a good's production. In the production approach, for example, the emissions associated with a cell phone produced in China would be recorded in China's national inventory, even if the phone is exported to the United States. In the consumption approach, the emissions generated by the phone's production would be assigned to the United States, where the phone is purchased and used. While a consumption approach may be more accurate in terms of understanding the demand driving emissions, it presents considerable implementation challenges if used for reporting and compliance.[28]

The question of embodied carbon is significant. China is the world's largest emitter; however, a nontrivial proportion of its emissions are generated in the production of exports rather than for domestic consumption. One estimate indicates that China's territorial emissions are roughly 9 percent lower than consumptive emissions, indicating the importance of embodied carbon in interpreting carbon accounting.[29] And as discussed in chapter 3, the EU's Carbon Border Adjustment Mechanism is premised on calculating the embodied carbon of imported products.

Corporate Reporting

When Walmart, Google, or Volkswagen want to measure their emissions, they rely, either explicitly or indirectly, on the Greenhouse Gas Protocol, the ur-standard in organization-level GHG accounting.[30] Created in 2001, it has become the basis for all subsequent organizational-level measurement standards and practices.[31] It has even been formally enshrined as an international standard by the International Organization for Standardization.[32]

The GHG Protocol, created by two NGOs, divides organizational emissions into the three "scopes" described earlier. Scope 1 emissions are directly emitted by a firm or organization, say in the process of refining oil or generating electricity. Scope 2 emissions are generated by electricity consumption. And Scope 3 is basically everything else. Since Scope 3 emissions are generated along the supply chain by other firms outside of the reporting firm, compiling original data is challenging—if not impossible. For example, Walmart has over 100,000 suppliers generating its Scope 3 emissions.[33] Complete reporting of these emissions would involve collecting data for *every* one of the 100,000 suppliers scattered across the globe and then, ideally, having these data independently verified. This is a herculean task. Moreover, this model assumes that supply chains are linear, but that is often not the case.[34] Rather, modern supply chains are more likely to be networks, making modeling more complex.[35]

Estimating Scope 3 emissions requires a suite of decisions about how to set organizational boundaries. The GHG Protocol offers three different approaches for this complex process. Boundaries can be defined by equity share, financial control, or operational control.[36] This design, one scholar has argued, "results in corporate GHG inventory reporting boundaries that are ambiguous and theoretically without limit."[37] And implementing these accounting practices creates interesting challenges. As Madison Condon notes, "How do you divide up the emissions between the milk and the meat that a single cow produces over its lifetime?"[38]

The GHG Protocol also provides some leeway about which Scope 3 emissions are "required" (in quotations since all reporting is voluntary). The standard includes fifteen categories for Scope 3 ranging from emissions associated with investments to the use of sold products. And each of these categories has both minimum requirements and optional components. For instance, firms *must* report the emissions associated with the transportation and distribution of their products. Thus, Amazon has to include all the emissions from contracting delivery companies. However, it can choose whether to report the emissions generated by the manufacturing of those vehicles. In other words, Scope 3 emissions have their own Scope 3 emissions. It's turtles all the way down.

A suite of tools can be used to estimate various Scope 3 emissions in the many cases where original supplier data are unavailable. These databases calculate the embodied carbon for various products, based on financial or procurement data. Precisely because of the resource intensity of calculating site-specific emissions, many databases use country or industry averages, which clearly overlook important variation. For instance, all Vietnamese rice is assigned the same emissions, whether methane-reducing practices are used in its production or not.[39] These estimation problems are exacerbated by the challenge of updating the data. Some of the databases used to estimate Scope 3 emissions in various sectors have not been updated in a decade.[40]

Problems about boundaries and data availability are only some of the challenges created by net zero. Scope 3 has its origins in life cycle analysis (LCA), which was created primarily as a decision-making tool for individual products, not entire organizations. LCA was made to assess "the emissions and/or environmental impacts from the processes associated with the production, use, and disposal of a specific product."[41] LCA was not meant to be applied to entire organizations, nor over unspecified periods of time—which is essentially how the GHG Protocol has adapted it.[42]

And yet, in many sectors, estimating Scope 3 emissions can be critical to meaningful net zero pledges. For instance, services and manufacturing sectors tend to have very high Scope 3 emissions, especially when compared to direct emissions.[43]

An example helps illustrate why the inclusion of Scope 3 emissions is so critical to the integrity of many net zero pledges. These emissions are particularly important in the technology and automotive sectors. For example, 77 percent of Amazon's emissions and 95 percent of Apple's are Scope 3. In the auto industry, the figure is even higher, since the vast majority of emissions are associated with the use of the car by consumers (assuming an internal combustion engine). Ninety-six percent of all emissions from Stellantis products (including brands such as Chrysler, Jeep, and Fiat) occur in Scope 3. For Volkswagen, the figure is 91 percent.[44] Thus, a clear picture of net zero pledges requires an understanding of the extent to which firms are including Scope 3 emissions in their pledges.

For all the reasons explained here, Scope 3 emissions are generally considered the most problematic in terms of GHG accounting.[45] However, even Scope 2, which focuses primarily on a firm's electricity consumption, has been shown to create problems that cause firms to underestimate their emissions.

Firms can reduce their Scope 2 emissions by purchasing renewable energy certificates (RECs). The claim is that by paying for the use of renewables in lieu of fossil-based energy, the firm has reduced these emissions. Unfortunately, there is ample evidence that purchasing RECs does not necessarily correspond to fewer emissions.[46] As Michael Gillenwater notes, the purchase of RECs (and the use of other market mechanisms like offsets) shifts from physical GHG accounting to financial transactions.[47] In the former, entities measure the actual atmospheric emissions assigned to an actor over time. In the latter, entities instead estimate the "economic value of GHG-associated emissions and removals and/or impacts over time."[48] Thus, the purchase of RECs does not necessarily change the *actual* source of energy consumed by the firm, but true to policies that manage tons, it allows fossil asset owners to claim green behavior.

Who's Doing What: An Assessment of Net Zero Pledges

To understand whether net zero pledges are a first step toward serious action, cheap talk, or something in between, I turn now to work I've conducted with colleagues Tom Hale and Aldrick Arceo that delves more deeply into the timing and features of both government and firm net zero pledges. This work compiles historical data on the proliferation of net zero pledges among states and firms as well as the robustness of those pledges.[49] To our knowledge, it is the first comprehensive historical analysis of net zero.

Our research examines all states that are a party to the Paris Agreement and all Fortune 500 firms.[50] We track the quantity of net zero targets and the timing of their adoption. We then create an index to assess their robustness from 2015 to the present. Robustness seeks to capture whether the pledge follows best practices and is not an assessment of ambition. It is primarily based on best-practice criteria as defined by the UN Race to Zero campaign.[51]

This work yields three key findings. First, net zero may be an old idea, but it is a new policy. Our data point to a "wave" of net zero pledges since 2020. There was a significant lag between the adoption of the Paris Agreement—which invokes "balancing" as an organizing principle—and widespread adoption of net zero pledges among both firms and governments. This finding is important: net zero is pervasive in climate discourse, but it is an extremely new policy for most actors, particularly governments.

National net zero targets (ranging from proposals to laws) covered less than 10 percent of global GDP until 2020, the majority in the realm of soft law.[52] Even today, only about 25 percent of global GDP is covered by binding legislation, though nearly the entire world is covered by targets that range from proposal to fully executed legislation. The paucity of net zero pledges inscribed into law is critical: without legislative enforcement, they remain aspirational plans.

Second, and perhaps surprisingly, firms have been leaders in the net zero wave. Political science research has shown that firms adopt climate policy when they believe that it will postpone more onerous regulations and/or when they have room to influence the rules or their implementation.[53] Net zero appears to be another data point in this phenomenon. In the context of existential politics, fossil asset owners' willingness to adopt net zero pledges indicates that they do not see net zero as a fundamental threat to asset value.

The data show that firms have moved more quickly on net zero than governments. By 2020, just as the big surge in national pledges was beginning, 40 percent of the total revenue of Forbes 500 corporations was already covered by net zero pledges. This enthusiasm should raise doubts about the extent to which net zero actually constrains the power of fossil asset owners.

Third, the robustness of net zero pledges signals cause for concern. We assessed pledges according to whether they include details on eight different attributes: target status, target year, interim target, gasses covered, plan (that is, whether there are steps to move toward net zero), reporting, accountability, emissions scopes covered, and the use of offsets. Using these attributes, we created an index to assess the robustness of individual pledges.[54] This is not an evaluation of ambition, but rather of the extent to which pledges conform to best practice, as identified in both the scholarly and gray literature.[55]

For governments, net zero pledges increase in robustness over time but fall slightly as more countries adopt them. Thus, later pledges are less robust than earlier ones—despite the fact that there is now some high degree of convergence on best practices.[56] Corporate pledges consistently increase slightly in robustness over time, even as more firms pledge to go net zero, indicating that later joiners are not laggards. Moreover, firms' average robustness starts higher than that of states, and though there are some wobbles, it generally remains higher. This suggests that firms have a vested interest in maintaining the credibility of the net zero approach. Greenwashing postpones asset revaluation.

Examining the individual attributes of the pledges demonstrates important variation. For example, the number of firms and governments that include interim targets has grown over time. Given the importance of peaking emissions as soon as possible, it is critical that actors specify how much they aim to reduce emissions in the short term. This is an important sign of improvement.

However, other attributes of the robustness index create cause for concern. Accountability mechanisms are scant. For governments, accountability is operationalized as whether the government is held accountable by the legislature, the courts, or other bodies for achieving the target. For most years, fewer than 10 percent of government pledges contained an accountability mechanism. Thus, there is little consequence for non-achievement. For firms, accountability is measured by linking executive pay to net zero achievement. Accountability mechanisms are higher for firms, ranging between 10 and 50 percent of all pledges, but this number falls over time.

Offsets present even more of a problem. The majority of governments and firms fail to specify how they will use offsets or to include them in their pledges without any limits or conditions on their use. And research shows that companies in particular tend to buy cheap, low-quality offsets. One study estimates that 87 percent of credits purchased by firms carry a high risk of not providing real and additional emissions reductions.[57]

Given the myriad problems with offsets detailed in the previous chapter, this cavalier approach could seriously undermine the integrity of the pledge—even though most net zero guidelines call for limiting their use to longer-term residual emissions.[58] It also shows that the problems associated with managing tons are layered: problems with net zero are also problems with offsets. This creates even greater possibilities for inaccuracies and outright gaming.

Governing Net Zero

The growth in net zero pledges has been accompanied by a proliferation of rules and guidelines about how to "do" net zero. As both governments and firms continue to figure out implementation, they are creating a variety of new governance initiatives to guide their efforts. Since the Paris Agreement, both international organizations and NGOs have created a slew of new voluntary rules and guidelines about the *processes* through which net zero plans are created, implemented, monitored, and reported. In general, these efforts focus less on the *substance* of the plans, though, as mentioned earlier, best practices are emerging about what constitutes a robust net zero pledge.[59]

This dense landscape of governance initiatives is yet another example of the layered problem of managing tons. The proliferation of net zero rules and standards has given rise to a cottage industry of coordination efforts to ensure that these initiatives are aligned or at least do not conflict with each other. Coordination gives rise to more meetings and even "meta-standards"—net zero standards to govern net zero standards.[60] This self-referential loop of rulemaking keeps everyone busy while moving them ever further from the underlying challenges of existential politics.

A 2022 report counts thirty-three different standards now governing net zero.[61] Some, like the Greenhouse Gas Protocol and CDP (formerly the Carbon Disclosure Project), are entirely created by nonstate actors, mostly NGOs. These are forms of "private authority," exercised by nonstate actors making rules and setting standards in world politics that others adopt.[62] Others, like the UN Race to Zero, are partnerships between international organizations and nongovernmental organizations. There is now even a set of guidelines from the International Organization for Standardization (ISO) about how to mainstream net zero into all other standards.[63] For brevity, I refer to both types of net zero rulemaking as "soft governance."

The Science Based Targets initiative is at the hub of net zero soft governance.[64] A partnership between the United Nations Global Compact, several large international NGOs, and the international disclosure organization CDP, SBTi provides standards on how to develop a net zero plan as well as sector-specific guidance for different industries. Members must sign a commitment letter indicating their intention to develop a net zero plan and then provide a detailed plan within twenty-four months. The plan is reviewed by SBTi, which approves those it deems consistent with its criteria. As of November 2024, over 9,400 companies have SBTi-approved targets.

Another key voluntary initiative is the UN Race to Zero campaign. Launched at COP26 in 2021, it is a network of networks that mobilizes non-state actors to develop and implement robust net zero plans in line with achieving the Paris Agreement goal of limiting warming to 1.5 degrees Celsius. With over 13,000 members, including firms, subnational governments, financial institutions and NGOs, it has developed into a large organization with extensive governance infrastructure.

The Race to Zero campaign has five basic criteria—pledge, plan, proceed, publish, and persuade—as well as an "interpretation guide" for implementing each of these steps.[65] Members must publicly disclose their plan within a year of joining and report on which measures they have implemented by the end of the second year. Interestingly, members are required not only to implement their own net zero plans but also to persuade others to do so as well: They must "align" their outward-facing activities (such as lobbying, association membership, and advocacy) with the twin goals of halving emissions by 2030 and reaching global net zero by 2050.[66]

Race to Zero and the SBTi are among the most central net zero standards; virtually all other voluntary governance initiatives refer to one or the other in their own standard-setting documents.[67] Yet why do these actors have the authority to make rules, and how do others know they are legitimate? Global governance scholars have already thought extensively about these questions. Some important themes emerge from this work that are relevant to the emerging landscape of net zero governance.

Making "Good" Rules

Voluntary governance works only if rules are credible and legitimate. Thus, there is an emphasis on transparency and coordination across soft governance efforts to reduce fragmentation and to ensure the diffusion of best practices. Coordination, or "alignment," has become a hallmark of soft climate governance, both in the voluntary carbon market and now in the net zero policy landscape.

Alignment—essentially a form of mutual policing—aims to ensure that voluntary governance is meeting these informal requirements of credibility, legitimacy, and robustness. We saw this phenomenon in chapter 4 with offsets: initiatives such as the International Carbon Reduction and Offset Alliance and the Integrity Council for the Voluntary Carbon Market aim to ensure that different voluntary offset standards are meeting mutually recognized robustness criteria. The same process is underway in net zero governance through institutions like the International Organization for Standardization and the UN High Level Panel on the Net Zero Emissions Commitments of Non-State Entities, which I discuss in further detail in the following.

While in theory, these types of mutual policing should be helpful for safeguarding credibility, in practice they suffer from two main shortcomings. Mutual policing is subject to conflicts of interest when these initiatives provide the veneer of legitimacy—through transparency and broad consultation—but fail to actually improve the robustness of rules. This is why offset quality initiatives reconvene in new forms every few years after failing to substantially address problems.

Moreover, the exercise of alignment pushes actors even further away from existential politics. Actors involved in soft governance hold meetings and consultations, release white papers, and draft *more* rules to ensure the quality and credibility of all the voluntary governance activity. Rules beget rules. This work means that everyone can busy themselves making rules about rules instead of dealing with the elephant in the room. I explore these political problems further in the next section.

The Political Challenges of Net Zero

Now that we have a better handle on the technical difficulties of carbon accounting and the complex landscape of net zero governance, the political challenges of net zero come into clear focus.

The clearest problem is greenwashing. The Oil Sands Pathways Alliance, a coalition of six oil and gas companies responsible for producing 95 percent of Canada's dirtiest fossil fuels, has pledged to go net zero by 2050. However, a recent study of the Alliance's advertising and public communication identifies problems of "selective disclosure and omission, misalignment of claim and action, displacement of responsibility, non-credible claims, specious comparisons, nonstandard accounting, and inadequate reporting."[68]

Relatedly, rules vary widely in their quality. And standards can be more or less rigorous in non-obvious ways. Taken together, this means that widely adopted standards are not necessarily the most robust or credible, as the story of SBTi illustrates.

The Science Based Targets initiative is one of the most widely used sets of net zero guidelines and is often invoked by other voluntary standards as a basis for their own rules. However, it has wrestled with both technical and political problems. On the technical side, SBTi does not specify a baseline year (only stating that it must be after 2014), even though the choice of baseline has a direct impact on the required emissions reductions. This flexibility allows firms to choose opportunistically, selecting later years, or those with higher emissions. This choice can make meeting interim targets easier. A report by the NewClimate Institute and Carbon Market Watch notes that "some companies . . . set targets compared to a base level of emissions that may in reality require hardly any further emission reductions between 2019 or 2020 and the target date."[69] This is not just a theoretical concern. Alternative calculations by NewClimate and a number of scholars suggest that the SBTi's methodologies are not in fact consistent with the Paris 1.5 degree target.[70]

SBTi has had political problems as well.[71] Until late 2023, firms paid SBTi to approve their targets. This clear conflict of interest was even raised by a former member of the initiative's Technical Advisory Group.[72] Eventually, demand for pledges outstripped SBTI's ability to manage the influx: there were simply too many firms seeking approval, and the team tasked with verifying these requests was too small. The combination of these two factors prompted a restructuring at SBTi. Pledges are now validated by an external group.[73] And SBTi has created a new independent Technical Council to strengthen standards to ensure that "its technical governance and standard-setting processes [are] in line with internationally recognized best-practice."[74] It remains to be seen whether and how these changes will address political and implementation challenges.

Despite these concerns, SBTi is continually cited by other guidelines and political processes around net zero. The UN's High-Level Expert Group on the Net Zero Emissions Commitments of Non-State Entities (HLEG), which aims to reduce greenwashing, refers to SBTi as an example of a "robust methodology" and a "best practice" that should be used by nonstate actors in creating a 1.5 degree–aligned net zero pledge.[75]

Accountability is another big political problem for net zero governance. The paucity of accountability mechanisms makes greenwashing—by both states and firms—much easier. Just 28twenty-eight countries have put their net zero promises into binding laws.[76] While it is true that many countries have pledged to go net zero through their nationally determined contributions (NDCs) under the Paris Agreement, these plans are not legally binding. Only the procedural parts of the Paris Agreement—which require new NDCs every five years and the ratcheting up of NDC ambition—are legally binding.[77] These provisions go some way to ensuring accountability, but they are not the same as a legal consequence for failure to deliver on net zero promises.

On the corporate side, roughly 30 percent of pledges have an accountability measure. To reduce greenwashing, some governments are starting to regulate firms' net zero claims more carefully. In the United States, the Securities and Exchange Commission (SEC) adopted a rule requiring firms to disclose their emissions, including Scope 3 emissions, when they are financially relevant or if the firm has a public emissions target.[78] Other countries have adopted rules around financial product standards, climate risk disclosure, decarbonization-focused procurement, and transition plans.[79]

At the international level, the UN High-Level Expert Group has also tried to rein in false or misleading net zero claims. Created in 2022, its goal is to create "a roadmap to prevent net zero from being undermined by false claims, ambiguity and greenwash."[80] It provides ten recommendations to enhance the quality of net zero pledges, while also noting that regulatory requirements are a superior route to achieving that goal.

Like other accountability mechanisms, however, the HLEG is caught in the self-referential loop of "alignment." In its very first recommendation, it calls for firms to align their pledges with robust (voluntary) methodologies such as SBTi, the Transition Pathway Initiative, or the International Organization for Standardization.[81] In other words, in keeping with the approach of managing tons, the HLEG aims to maintain consistency across measurement approaches.

To be fair, the HLEG does address the phaseout of fossil fuels (recommendation 5), calling for an end to development of new sources of coal, oil, and gas consistent with a 1.5 degree target. Other international organizations have made the same claim more strongly, though with little effect on state and firm behavior. Interestingly, there are no standards or even accepted benchmarks around phaseouts, so there is nothing to align.

This self-referential loop has reverberating consequences. When voluntary standards are coordinated and "leaders" such as the SBTi emerge, they are perceived as legitimate and sufficiently robust by governments. Then, as governments craft their own regulations, they base their rulemaking on these private standards. Just as in the voluntary offset market, these voluntary rules then "harden" into regulations.[82]

A handful of voluntary net zero rules are being adopted by governments in their own regulations on issues ranging from net zero–related claims, climate risk disclosure, procurement practices, and transition plan requirements.[83]

The export of voluntary guidelines into binding government regulations makes issues about credibility, legitimacy, and accountability even more important. Governments risk transforming low-quality voluntary rules into binding law. If problematic voluntary rules become law, firms and governments will be able to claim policy successes even when they are effectively climate failures.

Since net zero is still a new field of governance, rule hardening is also still emerging, as shown by some important examples. In 2023, the US Securities and Exchange Commission implemented a rule to standardize climate risk disclosure, basing reporting on the GHG Protocol.[84] Another proposed rule would require major suppliers to the federal government to disclose both climate emissions and risks. This rule also invokes private net zero standards, including the GHG Protocol, SBTi, CDP, and the Task Force on Climate-Related Financial Disclosures.[85] A similar process is underway with the European Union's regulation on corporate sustainability reporting, which also relies on the GHG Protocol.[86] These examples are part of a broader trend of governments incorporating voluntary standards into national regulations. One report finds that roughly 40 percent of governmental net zero standards rely in some way on voluntary standards.[87]

We have seen this playbook before. Both the international aviation agreement (CORSIA) and the Article 6 markets accept offsets from the voluntary carbon market—another version of private authority.[88] And the California emissions trading scheme accepts offsets based on methodologies developed by the Climate Action Reserve—another voluntary offset standard.[89] Elsewhere in environmental governance, private standards around forestry sustainability have been incorporated into EU timber regulations[90] and the 2008 US Lacey Act amendments regulating the import of forestry products.[91]

My colleague Tom Hale has argued that this hardening can be a source of robust regulations. He describes a "conveyor belt" model, whereby strong voluntary standards are refined, scaled up, and eventually incorporated into regulations. However, he notes, separating strong voluntary rules from weak ones, through UN-based initiatives like the HLEG, will be critical to ensuring that domestic regulations incorporate robust and stringent forms of private authority.[92]

The widespread recognition of SBTi as a key benchmark for net zero alignment is an initial indication that these alignment processes are not, in fact, sorting the wheat from the chaff. The experience from the voluntary offset market also provides reasons to question whether this sorting process will result in the hardening of quality standards rather than the weak standards that simply further legitimate managing tons.

Preliminary evidence from SBTi does not bode well for the hardening of quality standards. In April 2024, SBTi's board of trustees decided that firms could expand the use offsets for Scope 3 emissions. The decision provoked outcry from SBTi staff, who called for the CEO's resignation in response.[93] Around the same time, SBTi began work on revising and updating its corporate standard, which will come into effect in 2026. The revisions will address issues around Scope 3 emissions and "enhance interoperability with other SBTi standards as well as other relevant external frameworks and standards."[94]

An Elaborate Distraction

Even if one were to accept the critiques I have leveled here, net zero could still be rescued as a credible approach *if firms used pledges to tackle the problem of fossil fuel supply directly.*

Indeed, the largest political challenge of net zero is not the quality of the rules or whether they are aligned with each other or verified by third parties, but rather the paucity of pledges—by both states and firms—that directly address the problem of supply.[95]

It is no secret that we are rapidly blowing through our remaining carbon "budget"—the amount of GHGs we can emit while still achieving the Paris temperature goal.[96] This means that curtailing the supply of fossil fuels is equally as important as slowing demand, if not more so.[97] (This is why activists and some of the most vulnerable island states have called for a Fossil Fuel Non-Proliferation Treaty.[98])

And yet the phaseout of fossil fuels is noticeably absent from most net zero pledges. In a 2023 report, the Net Zero Tracker found that only 6 percent of countries and 4 percent of firms have committed to a full or partial phaseout of oil exploration.[99] The figures for oil production are not much better: 7 percent of countries and 23 percent of firms have committed to a full or partial phaseout, though only 3 percent of firms have committed to a complete phaseout. Natural gas presents a similar story. Countries in particular are more ambitious when it comes to curbing coal use; nevertheless, 77 percent of countries have no stated plans to phase out coal.

Net Zero Tracker's work is consistent with my own research with colleagues on the top ten investor-owned oil companies. We examined their political behavior, which we measured by coding investors' earnings calls between 2004 and 2019. Though firms increasingly accept the science of climate change and the need for international rules and a carbon price, they consistently deny that fossil fuels must be phased out, across all years of the sample and between 2008 and 2016 in particular. We find that no firm has publicly supported a fully decarbonized economy.[100]

The problems with the implementation of net zero are formidable. Ultimately, they distract from fossil asset owners' bigger problem: asset revaluation. Yes, net zero can help states and firms figure out a plan to get off fossil fuels. But thus far, there is little indication from the pledges themselves or from the rising concentrations of GHG emissions that either group has made meaningful efforts to tackle supply. Until this happens, net zero will remain an elaborate distraction.

PART III

Focusing on Assets

6

Hit 'em Where It Hurts

CONSTRAINING FOSSIL ASSET OWNERS

THIS IS THE FIRST of two "solution" chapters. I've argued that managing tons is not an effective strategy for decarbonization. Existential politics means that governments cannot simply invest in green assets, they must also constrain the power of fossil asset owners. In this chapter, I tackle the difficult question of how to rein in fossil asset owners. This will not happen through the UNFCCC, which is primarily interested in managing tons, but rather through international institutions that engage directly with asset revaluation: taxation and investment regimes.

Constraining fossil asset owners has three important benefits. First, it is well documented that fossil asset owners have spent decades and billions of dollars to preserve the fossil fuel economy.[1] Removing their financial advantages will help curb their ability to obstruct progress and will also hit fossil asset owners where it hurts—on their bottom line.

Second, reforming tax and finance institutions frees up money that could be used for green assets and the energy transition more broadly. Of course, governments will make their own decisions about how to spend any money they may recover through changes to these institutions; there is no guarantee that it will be invested in green asset owners. But *even if recouped funds are not spent on decarbonization*, these reforms would still reduce the material wealth of fossil asset owners—which is a win for the climate and for climate politics.

Finally, the policies discussed in this chapter can help reduce wealth inequality, which is a key cause of climate change.[2] In 2019, the richest 1 percent of the planet emitted as much as the poorest 66 percent—which is five billion people.[3] And the average billionaire emits three million tons of CO_2e per year through their investments alone.[4] Taxing the multinational corporations that are the source of this investment income will help reduce the growing gap between rich and poor.

A key assumption underpinning both solution chapters is that the current liberal international order—the multilateral institutions that facilitate open markets and create interdependent economies—isn't going away. Of course, we must acknowledge the extensive problems with, and injustices created by, these institutions. But for the time being, governments have to work within these constraints. Some have argued that it is "capitalism versus the climate,"[5] but we simply do not have time to replace capitalism and its international institutions with something new. However, governments can radically reform the rules that govern them. These reforms can both speed the energy transition and create more just societies.

Thus, these two chapters are written from what I would call (perhaps unsatisfactorily) a position of "radical pragmatism." Radical pragmatism gets at the root of the climate crisis by seeking to change the balance of fossil and green assets in the global economy rather than managing tons. And it offers a path for significant departure from the status quo through the redistribution of capital and reassertion of state sovereignty in an age of globalization.

This chapter describes the first step in imagining global climate politics beyond the UNFCCC by detailing changes to regimes that manage capital that could promote decarbonization by shrinking the economic power of fossil asset owners. These are not new ideas; others have called for reforms of both the tax and investment regimes.[6] The key contribution is shifting the mental model from managing tons to focusing on assets. Some of these changes may not seem radical, since they move at the measured pace of multilateralism. But as I explain, they have the potential to set larger changes in motion.

I first discuss the concept of radical pragmatism and how it can help address the conflicts created by existential politics. I then turn to a discussion of two institutions that can help constrain the material power of fossil asset owners: the OECD's new rules on a corporate minimum tax, and the slow but steady chipping away of fossil fuel protections in the Investor-State Dispute Settlement (ISDS) system. The former represents a profound shift in state sovereignty. Taxation is a core function of the state, yet the new rules require signatory states to agree to standardize their tax rates at a minimum of 15 percent. The latter is more incremental in its progress but could reconfigure the costs and benefits of investments in fossil fuels.

Radical Pragmatism

Radical pragmatism acknowledges two conflicting realities. First, addressing the climate crisis will require big changes to the current international order. And second, the economic and political dominance of fossil asset owners severely constrains governments' ability (and in many cases, their desire) to make those changes. Radical pragmatism tries to square this circle.

The reforms I discuss in this chapter are radical not only because they address the political conflicts created by asset revaluation, but also because they could catalyze even greater change. They are pragmatic in the sense that they operate in the realm of the possible: there is already considerable momentum for reform in both the tax and investment regimes.[7]

These reforms are incomplete, to be sure, but they are more likely to generate real decarbonization than managing tons. Achieving deep decarbonization will require undoing "carbon lock-in"—the intersecting economic, social, and technical systems that produce our reliance on fossil fuels.[8] Managing tons does little to destabilize these mutually reinforcing systems. By contrast, raising fossil asset owners' corporate taxes and revoking their investment protections begins to chip away at their economic dominance. And even though obstructionism is a huge challenge for climate policy, these reforms operate squarely in the realm of the possible.

Radical pragmatism also requires acknowledging what won't work. I do not address the obvious policy that would help reduce the material power of fossil asset owners: subsidies. Globally, governments provided $7 trillion in fossil fuel subsidies in 2022.[9] Stopping the flow of free money to these fossil asset owners would reduce emissions in 2030 by an estimated 43 percent below business-as-usual levels, keeping the world on track to meet the Paris temperature goals.[10] It would also hasten the asset revaluation process.

Eliminating fossil fuel subsidies is a no-brainer, but politically it's a nonstarter. Since first resolving to phase out fossil fuel subsidies in 2009, the G20 has consistently failed to deliver on that promise. Unlike the other reforms examined in this chapter, subsidy reform has not gathered any real political momentum—beyond empty rhetoric—for almost two decades.[11]

Tax Reform Is Climate Policy

Correcting tax evasion is a necessary but not sufficient condition for ambitious climate policy.

TABLE 6.1: Top Tax Havens, 2021

1. British Virgin Islands	9. Singapore
2. Cayman Islands	10. United Arab Emirates
3. Bermuda	11. Ireland
4. Netherlands	12. Bahamas
5. Switzerland	13. United Kingdom
6. Luxembourg	14. Cyprus
7. Hong Kong	15. Mauritius
8. Jersey	

Source: Corporate Tax Haven Index 2024.

Recent scholarship shows that multinational corporations (MNCs) evade or avoid astronomical amounts of taxes through "offshoring." Offshoring is, quite simply, regulatory arbitrage: MNCs report (or book) their profits in tax havens: jurisdictions with low or zero tax rates, many of which are small island nations like the Cayman Islands. However, many European nations also have low corporate tax rates as a strategy to attract corporate business. Table 6.1 lists the top tax havens according to the watchdog NGO Tax Justice Network.[12]

Tax havens are not new. Switzerland was the first nation to become a tax haven, in the 1920s.[13] But they did not begin to proliferate until the 1980s, when new offshore tax havens emerged in places like the Bahamas, Jersey, Luxembourg, and Singapore. They have expanded over the last four decades, and correspondingly, corporate tax rates have fallen precipitously, particularly in the last twenty years. As a consequence, countries have suffered huge losses in revenue.

The numbers on offshoring are nothing short of astonishing. Estimates suggest that anywhere from $7 trillion[14] to $32 trillion[15] in assets are in offshore accounts, where little to no tax is paid. Moreover, over one-third of corporations' multinational profits are "offshored" to avoid some level of taxation, amounting to some *$1 trillion.*[16] US firms are particularly active in offshoring, shifting twice as much profit as other multinationals relative to their foreign earnings.

Offshoring creates huge revenue losses for governments. Studies suggest that global revenue losses range between $500 billion[17] and $850 billion annually.[18] *The Guardian* reports that, in 2021 alone, the United Kingdom lost more than £2 billion due to offshoring by big tech companies.[19] In a world without this shifting, domestic profits would increase by about 20 percent in the European Union, 10 percent in the United States, and 5 percent in developing countries, and they would fall by 55 percent in tax havens.[20]

How Tax Policy Helps Fossil Asset Owners

Many types of multinationals engage in tax avoidance. Here I focus on emissions-intensive fossil asset owners, such as oil, gas, and mining companies. It is difficult to find systematic data on the extent of the practice in the fossil fuel industry, but scholarly papers and reports from the gray literature indicate that tax avoidance is a significant problem.

Developing countries, who have long decried the unfairness of the current global taxation regime, are disproportionately affected by tax avoidance and profit-shifting.[21] In sub-Saharan Africa, the economies of fifteen countries are heavily reliant on mining, and ten of them are among the top fifteen economies most reliant on mining globally.[22] Mining contributes almost 9 percent of global GDP and more than half of total exports for the average resource-rich sub-Saharan African nation.[23] Much of this income comes from mining multinational corporations, such as Glencore Xstrata, Rio Tinto, and Anglo American, which provide foreign direct investment for extraction and pay tax on these economic activities. Yet a World Bank study finds that profit-shifting in the mining industry results in annual lost tax revenue of between $470 million and $730 million.[24] Other work finds that the European oil and gas firms Total and ENI are set to avoid $2 billion in taxes in Mozambique by routing profits through the United Arab Emirates on two natural gas projects.[25]

Fossil asset owners are also avoiding taxation in advanced economies. The Tax Cuts and Jobs Act that went into effect in the United States in 2018 includes a provision for a minimum tax on profits earned abroad but allows generous exemptions for oil and gas firms' foreign profits.[26] A 2021 report by three international NGOs calculates that these carve-outs are worth $86 billion.[27]

North of the border, the NGO Canadians for Tax Fairness finds that major Canadian oil companies, such as Suncor, Enbridge, CNRL, TransCanada, Imperial Oil, Cenovus, and Husky, all have subsidiaries in tax havens that they use to lower their domestic tax burdens. They estimate that these tax avoidance practices cost the Canadian government between CAD$10 billion and $15 billion per year.[28] And of course, this is on top of the approximately CAD$20 billion in subsidies that Canada provides to the oil and gas industry.[29]

Australia has experienced similar avoidance problems with Chevron and BHP Billiton but has taken legal action to recoup its losses. In 2015, an Australian court found that Chevron used tax havens to avoid paying millions of

dollars and ordered the company to pay $654 million to the Australian govern-
ment to compensate for the lost tax revenues. A similar case was filed against
BHP Billiton, which was forced to pay more than half a billion dollars in taxes
avoided between 2003 and 2018.[30]

In addition to depriving states of revenue, offshoring is harmful in several
other ways. First, it reduces the sovereign power of the state. Recapturing missing capital is in part about finding monies to fund decarbonization, but more
importantly, it is about reasserting the role of the state in generating and ap-
propriating tax revenues. Low-income and lower-middle-income countries
are particularly affected by corporate tax evasion.[31] Thus, addressing the problem will be especially beneficial to these nations. If states do not intervene to
perform the basic function of taxation, it is not likely that they will intervene
at the scale necessary to aggressively decarbonize.

Second, offshoring by both individuals and corporations contributes to
income inequality, which is a key driver of climate change. In 2019, the bottom
half of the world's population were responsible for just 12 percent of global
emissions, whereas the top 10 percent emitted 48 percent of the total.[32] And
between 1990 and 2019, the top 1 percent were responsible for nearly one-
quarter of the growth in global emissions.[33] Offshoring allows for further
wealth concentration, and therefore more emissions.

Oxfam estimates that the richest 1 percent will have to reduce their carbon
footprint by a factor of at least thirty to meet the Paris Agreement target of
limiting warming to 1.5 degrees Celsius.[34] It is clear that reducing extreme
wealth and income inequality—which is exacerbated by tax avoidance via
offshoring—will help address the climate crisis.[35]

Third, offshoring undermines democracy. Taxes and political representation are inextricably linked, and when they are decoupled, demands for
political accountability diminish.[36] Moreover, when corporations amass great
wealth, they can use these resources to influence elections and regulation. Fossil asset owners have a long history of obstructionism, influencing politics and
policy to preserve the value of their assets.[37]

Finally, offshoring has been linked to climate crimes. Deforestation and
forest degradation are both huge contributors to climate change and often
happen illegally.[38] Beef and soy producers clear Amazonian rainforest to raise
cattle and crops, and they use tax havens to maximize profits.[39] Those profits,
in turn, allow them to continue deforesting. In Indonesia, an anonymously
owned forestry company has cleared more than thirty-three thousand hectares
in West Kalimantan since 2021, converting it to monoculture pulpwood

plantations. The company is owned by firms incorporated in Samoa, the Cayman Islands, and the British Virgin Islands, which do not require disclosure about "beneficial owners"—that is, the parties who own, control, or benefit from the company.[40] Once again, tax havens allow these fossil asset owners to continue activities that accelerate climate change.

The New Global Minimum Corporate Tax

Meaningful tax reform could help reduce tax evasion and avoidance, raise considerable revenue, reaffirm states' sovereignty, and reduce some of the material power of fossil asset owners. The good news is that such a process is currently underway. The OECD Global Anti-Base Erosion Model Rules (or simply, the Model Rules) impose a minimum corporate tax of 15 percent on corporations with more than €750 million in annual revenue. More than 135 countries have agreed to join, including the United States and the European Union. The Model Rules must be implemented through national legislation.[41]

Countries adopting the Model Rules agree to implement the minimum corporate tax of 15 percent. The tax will be levied on a *jurisdictional* basis—that is, where economic value is created.[42] This is a departure from a century of tax law, which tended to tax primarily in the "residence" country, even if the revenue was generated elsewhere.[43] Now, if firms operate in jurisdictions with a lower tax rate, they must calculate the difference between their effective tax rate and the 15 percent. That difference—or the "top-up"—is then owed in the jurisdiction of the parent of the MNC.[44] If the parent company operates through subsidiaries in low-tax jurisdictions, a "backstop" kicks in to increase the tax at the subsidiary level. The backstop is calculated in proportion to the relative share of assets and employees in the subsidiary.

The OECD Model Rules for international taxation represent an enormous shift in global financial governance, with important implications for the climate crisis. In the emerging regime—which now covers 90 percent of global GDP—countries have relinquished a key characteristic of sovereign states: the ability to set taxation rates. They have not only collectively agreed to an absolute rate but are reversing the decades-long practice of state competition for MNC tax revenue.[45]

The top-up provision (officially named the Undertaxed Payments Rule) ensures that if enough of the largest economies participate, the regime is effectively self-policing. There will be no advantage to shifting to lower tax jurisdictions, because they will also be subject to the top-up. Moreover, the top-up helps solve the nonjoiner problem: because their firms will be subject to the top-up, countries that do not join will miss out on the revenue gained by participating in the regime. In other words, staying out of the Model Rules will be a lose-lose for the nonjoiners. Any country that doesn't have a 15 percent tax rate will have the difference imposed by other countries where the firm has a presence. So the agreement will require coordination across countries to calculate and distribute the top-up appropriately for firms that operate in multiple jurisdictions.

As always, the devil is in the details—which reveal some of the weaknesses of the Model Rules. "Carve-outs" are permitted, so that certain parts of revenue are not subject to the minimum tax.[46] Although the carve-outs will be reduced over time, current estimates indicate that they will reduce total revenues from the minimum corporate tax by about 22 percent (across a sample of eighty-three countries).[47] These carve-outs allow countries to protect some of their domestic tax incentives where there is "substantive" economic activity. In other words, the carve-outs lessen the blow of the minimum tax. In response to these weaknesses, some have called for raising the minimum rate to 25 percent and for a comprehensive registry of all global wealth and assets.[48]

The Next Phase of Tax Reform

Now that countries have signed on to the Model Rules, the next step is enacting domestic legislation to implement its goals. Such legislation has been passed in the European Union, Japan, South Korea, and the United Kingdom. Draft legislation is underway in most EU and EU-adjacent countries as well as New Zealand.[49] Some tax havens, such as the Bahamas and Bermuda, are still considering various forms of a corporate income tax.

Despite these positive developments, there are some challenges ahead. The first challenge is the United States, which has signed the OECD framework but has different legislation in place. The Tax Cuts and Jobs Act created a Global Intangible Low-Taxed Income Tax (GILTI), which aims to capture tax revenue from some of the profits shifted abroad, though the rate is slightly different from the Model Rules rate. And the Inflation Reduction Act created a Corporate Alternative Minimum Tax (CAMT), which imposes a 15 percent minimum tax on large corporations; however, the CAMT does not comply with the Model Rules, for a variety of highly technical reasons.[50] The expectation is that these inconsistencies will be corrected through future legislation, but congressional agreement on this matter is far from guaranteed.

Because of this US exceptionalism, countries have agreed to provide a "transitional safe harbour" protecting US businesses from the top-up until 2026.[51] If the United States were to continue with its alternative approach, US businesses would technically be subject to the top-up. However, it is an open political question whether other governments want to take on US firms to enforce this rule.

Second, as with all international regulation, the Model Rules will only be as successful as domestic legislation to implement it. And implementation is a highly complex endeavor. The OECD has released four "tranches" of administrative guidance on how to implement the Model Rules, amounting to hundreds of pages. This guidance is not binding, nor is it subject to legal oversight.[52] Moreover, putting the OECD in the role of a (soft) rulemaker is a double-edged sword. It allows for insulation from influence at the domestic level, but without democratic safeguards[53]—an oft-cited critique of global governance institutions.[54]

Third, there is yet another multilateral process for tax reform underway: the United Nations Framework Convention on International Tax Cooperation. This proposed treaty is in its very early stages; governments have not yet begun negotiating it. Rather, they have met twice to craft the treaty's terms of reference, which will define the objective, principles, and substantive foci of the Convention.[55] The most recent draft (at the time of writing) overlaps considerably with the OECD Model Rules and proposes that the new treaty contain rules to govern multinational corporations and digital services.[56]

The proposed UN tax treaty appears to be a case of "forum shopping"— where some countries seek an international forum more favorable to their interests.[57] Indeed, many developing countries have voiced objections to the OECD process, which has long been critiqued as a club for wealthy countries. These objections are buttressed by evidence that the Model Rules will not benefit developing countries as much as their developed counterparts.[58]

The United Nations, by contrast, has universal membership and accords each country an equal vote. In part because of this egalitarian approach, some wealthy countries—including Australia, Canada, Israel, Japan, Korea, the United Kingdom, and the United States—have objected to pursuing a UN tax treaty.[59] It is unclear what the effect of the new treaty will be, both because the rules have yet to be determined and because of its relationship to the Model Rules. New forums can empower previously disenfranchised actors, but competition can also result in weaker rules.

For some, the new corporate minimum tax is nothing short of a revolution in international tax cooperation, with potentially huge implications for climate policy.[60] The Model Rules are poised to allow countries to reassert their sovereignty by recouping lost tax dollars. However, the creation of these rules is just the first step. To realize their full potential, countries must pass implementing legislation, in line with the administrative guidance put forth by the OECD. Moreover, they must sort out the relationship between the OECD's rules and the emerging UN Framework Convention on International Tax Coordination. Even with the challenges of these efforts, a sea change is clearly underway around norms and practices on taxation.

Investment Reform Is Climate Policy

The global investment regime is another area where revised rules could have profound implications for climate policy. Revoking the current set of investment protections for fossil fuel projects attacks the critical, and underregulated, problem of restricting supply by making fossil fuel extraction riskier and more expensive.[61] Ultimately, reforming investment rules could not only change the calculus of fossil asset owners and would-be fossil asset owners in developing new fossil fuel supply, but also deprive them of millions of dollars in future compensation.

The Investor-State Dispute Settlement is an international arbitration system that serves as the enforcement mechanism for investment treaties. When a firm undertakes economic activities in a foreign country (foreign direct investment), it seeks assurances from the host nation that its investments will be legally protected from threats such as expropriation, trade discrimination, or disproportionate regulatory burdens.[62] These protections are inscribed in the international investment agreement between two (or more) nations. Purported violations of those agreements can be arbitrated by the ISDS. Cases are brought by an investor firm against a state and decided by a panel of arbitrators appointed by both parties (one appointed by each and one selected jointly or appointed by an arbitral institution).

There are currently over 2,600 bilateral and multilateral investment treaties in force that provide access to the ISDS, including the Energy Charter Treaty (ECT).[63] Signed in 1994, the ECT was created to facilitate foreign direct investment, primarily to expand the energy supply for the global economy. Fifty-three countries are party to the treaty, though as I discuss later, a number of countries have withdrawn or are planning to do so.

Seen through the lens of asset revaluation, international investment law is climate policy.[64] It provides broad protections for fossil asset owners, creating "far-reaching rights" for investors, with minimal obligations.[65] Coupled with the opacity of the arbitration process and the reliance on private lawyers, the system is heavily weighted toward investors,[66] effectively dissuading aggressive state regulation of industry for fear of litigation by foreign firms.[67]

Most investment treaties contain provisions that constrain states' ability to pursue decarbonization policies. For example, firms can sue host countries if state regulations "substantially" affect the value of their investments, considering both the "legitimate expectations" of the investor and the proportionality of the regulations.[68]

In 2015, the UK oil and gas company Rockhopper Exploration sued the Italian government under the Energy Charter Treaty for its decision to ban offshore oil and gas exploration. Italy had granted Rockhopper a permit to explore an offshore oil field in 2005. However, in 2010 the Italian legislature banned all new oil and gas permits. The initial decision to grandfather in those permits issued prior to the law's passage was reversed, as a result of public and activist pressure. The government revoked Rockhopper's permits in 2012.

In response to the revocation, Rockhopper sued, arguing that the ban interfered with its investments in offshore exploration.[69] The tribunal sided with Rockhopper, ruling that Italy's decision to outlaw oil and gas exploration was tantamount to expropriation, and awarded Rockhopper €190 million in damages.[70] Following the ruling, the Rockhopper CEO announced that the monies would be used to invest in further oil exploration in the Falkland Islands.[71]

In an even wilder case, Australian billionaire and former politician Clive Palmer sued his own government for $300 billion in lost profits for a coal mine he owns in Queensland. Because the firm that owns the mine, Zeph Investments, is located in Singapore, it is covered by the investment protections in the ASEAN–Australia–New Zealand Free Trade Agreement.[72] Palmer initiated three separate cases seeking compensation for lost profits following Queensland's decision to prohibit exploration for the mine.[73] As of late 2024, the cases were still pending.

For the last three decades, the ISDS has protected fossil fuel companies, contributing to their material and political power. Since 2013, roughly 20 percent of ISDS cases involve the fossil fuel industry, and the average award (to firms) is $600 million.[74] Because of generous interpretations of the rules on calculating compensation, awards to fossil fuel companies have been particularly large compared to other sectors.[75] Eight of the eleven largest ISDS awards—all over $1 billion—have gone to fossil fuel companies.[76]

In an ironic development, some of the windfall energy taxes imposed by governments have been challenged under the ISDS. In 2023, a British oil refinery company, Klesch, challenged windfall energy taxes in Germany and Denmark, claiming that they were "arbitrary, discriminatory and punitive" and would result in a "substantial economic loss" for the company, putting it in a "substantially weaker competitive market position relative to unaffected companies."[77] The case is still pending, but a preliminary decision states that Klesch does not yet have to pay the tax, amounting to more than €95 million.[78] While some experts suggest that the case law does not establish a precedent for successful claims of indirect expropriation via windfall taxes, it appears that at least some fossil asset owners are willing to try their luck.[79]

These large sums are particularly challenging for developing nations. A long-standing dispute between Chevron and the government of Ecuador over concession contracts in the Amazon ended with a $70 million payout to one of the world's largest fossil fuel companies. A similar case in 2012 resulted in another $1.7 billion payout to Occidental. And since 2014, Venezuela has paid out three of the ten largest awards under the ISDS to ConocoPhillips, Mobil and the mining giant Crystallex, totaling more than $11 billion.[80]

Without significant reform to the ISDS, such lawsuits are likely to be brought more frequently and with larger price tags. A recent study examines oil and gas exploration permits that could be canceled owing to the increasing stringency of climate policy—as with the Rockhopper case. They find that the value of projects awaiting final permitting is between $60 billion and $234 billion (depending on projected oil prices).[81] Given that the IEA has stated that no new fossil fuel infrastructure should be built if we are to meet the Paris goal of limiting warming to 1.5 degrees Celsius, more cancellations are probable. And if current projects are added to the cancellations, the range of values for projects awaiting permitting rises to between $92 billion and $340 billion.[82] In other words, without either significant reform or withdrawal from ISDS, fossil asset owners will continue to reap handsome rewards from governments.

The ISDS both enriches and emboldens fossil asset owners. Fossil fuel companies need not win their cases; the mere threat of lawsuits is sufficient to create a "regulatory chill" that discourages countries from phasing out the use of fossil fuels.[83] Fear of (costly) arbitration may be enough to deter governments from taking stricter action. Indeed, government officials in Denmark, France, and New Zealand have conceded that they delayed oil and gas phase-outs for precisely this reason.[84] Nobel Laureate Joseph Stiglitz has a less demure term for governments' hesitancy to regulate: "litigation terrorism."[85]

Reforms Are Stalled, but Change Is Afoot

Cognizant of the perverse effects of the ISDS on climate policy, governments spent four years engaged in a process to "modernize" the ECT. Proposed reforms would have allowed signatories to exempt fossil fuel investments through a flexibility mechanism; narrowed definitions of key legal concepts, like expropriation and fair and equitable treatment; and made other procedural changes.[86] Advocates and scholars criticized many of these proposed reforms as insufficient to address the fundamental problems with the ISDS.[87] They failed to deal with critical issues like the long "sunset clause" that would protect investments for up to twenty years after countries exited the treaty and the extension of protections to controversial technologies like carbon capture, utilization, and storage.

However, debates about reform have taken a backseat to a growing movement of withdrawals. In 2024, the European Union and the United Kingdom each announced their withdrawal from the Energy Charter Treaty. This is good news for the climate: the European Union can effectively nullify the application of the sunset clause within its borders.[88] According to one study, this would eliminate 99 percent of all possible claims in the United Kingdom and 100 percent of such claims in the European Union.[89] Australia, France, Germany, Italy, Poland, and Russia have also already withdrawn from the ECT.[90]

The destabilization of the ECT is a promising sign that investment rules can constrain fossil asset owners. Other developments within the ISDS also provide reasons for optimism. The United States–Mexico–Canada Agreement (USMCA) governs trade between the three nations and entered into force in 2020. Under the terms of the treaty, the United States and Canada can no longer use the ISDS mechanism.[91] The exclusion of the ISDS was critical to protecting the United States from a $15 billion case brought by the Canadian firm TC Energy. The Biden administration revoked permits for the Keystone XL Pipeline, which would have allowed for the transport of Canadian oil to refineries in the United States. After TC Energy filed a claim against the US government, the ISDS tribunal ruled in favor of the United States. It noted that the revocation of permits occurred *after* NAFTA was replaced by the USMCA, which is not subject to ISDS provisions.[92]

A few countries have withdrawn from the International Center for Settlement of Investment Disputes, which serves as the institutional forum for ISDS arbitration. Bolivia, Ecuador, Venezuela, and, most recently, Honduras have all exited ICSID. Ecuador has a provision in its 2008 constitution (reconfirmed through a referendum in 2024) that prohibits the use of the ISDS. And Australia has stated that it will no longer agree to ISDS provisions in future investment agreements.[93]

Other efforts to align investment and climate goals are also underway. In 2021, the OECD began to further investigate how to align investment treaties with the goals of the Paris Agreement.[94] This process has yet to provide specific recommendations, though a series of workshops and stakeholder consultations have taken place. In late 2023, countries agreed to another two years of work, culminating in sample treaty language for limiting fossil fuel protections in investment treaties. Several participants in the OECD process are also members of the Beyond Oil and Gas Alliance (BOGA), a partnership of national and subnational governments working to phase out fossil fuel production. At least one nation (New Zealand) slowed its involvement with BOGA because of concerns about arbitration claims under the ISDS system.[95]

The international investment regime has made some adjustments to accelerate the devaluation of fossil assets. These adjustments are not enough, but they are a starting point.[96] Coupled with policies to catalyze the growth of green assets, which I discuss in the next chapter, efforts to devalue fossil assets can help shrink the power asymmetry between fossil and green asset owners.

Is Focusing on Assets Really Better than Managing Tons?

Skeptics might ask whether raising the minimum corporate tax or reforming the ISDS are actually better approaches than the UNFCCC's current approach of managing tons. Could these approaches not be captured by fossil asset owners as well? Aren't they similarly technocratic? Of course, any regulatory process can be captured, and these two regimes are no exception. However, reforming trade and investment rules has several advantages over managing tons.

First, both regimes get at the root of the problem of climate politics: money. Successful tax reform and revocation of investment protections for fossil fuel companies could make a meaningful dent in the wealth of fossil asset owners and therefore their ability to obstruct climate policy in a way that managing tons does not—even if provisions are watered down. *Any* additional tax income collected from fossil asset owners is money that cannot be spent on fossil fuel investments or obstructing climate policy. And even partial revocation of investment protections would substantially change fossil asset owners' calculus on the profitability of future investments.

This point bears repeating, since it gets to the crux of existential politics: *even if recouped funds are not spent on green assets, they can still help destabilize the political incumbency of fossil asset owners.* Radical pragmatism, even when incomplete, gets to the root of the issue: the gross economic and power asymmetry between fossil and green asset owners. The same is not true for managing tons; indeed, the opposite is true. As chapters 3 through 5 demonstrate, managing tons provides fossil asset owners with more leeway to continue with business as usual or to implement strategies like hedging and greenwashing.

Second, in the specific case of tax reform, fiat currency is much closer to a "real" commodity than GHG emissions. Managing tons is a relatively new phenomenon, one that requires extensive new governance capacity. By contrast, taxation is a core state function, and states have lengthy experience in levying and collecting taxes. As a result, although firms will undoubtedly exploit loopholes in the tax and investment regimes and try to game the new rules, governments are already well versed in this playbook. Thus, states have a leg up on detecting noncompliance or gaming, in comparison to the commodification and trade of GHGs.

Third, tax reform imposes a minimum standard (the 15 percent tax rate) rather than allowing states to set their own goals—which is the model of the nationally determined contributions in the Paris Agreement. This minimum standard has political advantages, even with the loopholes and workarounds noted earlier. It has transformed a cooperation problem (a race to the bottom through tax competition) into a coordination problem.

Thus, the main issue now is how countries will change their laws to implement the minimum tax. This shift can create pressure for international harmonization. If most jurisdictions use the new Model Rules, fossil asset owners incur transaction costs when they operate in countries that use different standards. Counterintuitively, this patchwork creates an incentive for fossil asset owners to push nonconforming jurisdictions to change their rules. This process of upward harmonization has been dubbed the "California effect." California has traditionally had stricter air pollution standards than the rest of the United States, and it requires car manufacturers to adhere to these standards. Given California's dominance in the US car market, automakers decided that it was easier and more cost-effective to have all cars manufactured to adhere to the higher standards rather than create two different types of cars.[97]

Fourth, tax reform is being implemented at a geopolitical moment when the doctrine of unfettered free trade is being questioned.[98] Rising wealth inequality and the erosion of the social contract are giving rise to progressive and populist movements, some of which threaten political stability. This pressure has provided additional political support for taxing MNCs and, in particular, for imposing windfall taxes on fossil fuel companies. For example, in 2022, under the Conservative leadership of Prime Minister Rishi Sunak, the United Kingdom imposed the Energy (Oil and Gas) Profits Levy of 25 percent on oil and gas companies operating in the country in response to the huge profits earned from price spikes resulting from the war in Ukraine. The levy has since been increased to 35 percent and is set to last until 2030.[99]

In a word, there is a robust societal demand for greater equality, which is much more legible to the public than "better" offsets, net zero pledges, or carbon pricing. And this political demand is an essential ingredient for producing global regulation in the common interest, like promoting rapid decarbonization.[100]

Finally, and perhaps most importantly, while the UNFCCC continues to flounder, there is real change happening in the area of tax reform as well as potentially important destabilization of the Energy Charter Treaty. There is ample momentum to build on in both areas.

Since climate change is, at root, a problem of asset revaluation, reining in fossil asset owners is a critical part of accelerating decarbonization. Reforming tax and investment regimes can help reduce the profits of fossil asset owners and the profitability of generating more supply. More importantly, reducing fossil asset owners' material resources can lessen the power asymmetry between fossil and green asset owners by weakening them politically. I now turn to the next task: considering ways to grow the strength and power of green asset owners through reforming the trade regime.

7

Green Industrial Policy

CREATING GREEN ASSET OWNERS

EXISTENTIAL POLITICS indicates two main pathways to promote rapid decarbonization: constraining the power of fossil asset owners and creating
green asset owners to advocate for more ambitious climate policies. This chapter addresses the second pathway by considering the critical question of how
to build green assets at the speed and scale that the climate crisis requires. As
in the previous chapter, the discussion embraces radical pragmatism—
acknowledging that the international institutions of the liberal order are
deeply flawed, but that the urgency of the climate crisis requires working
within the constraints of these institutions.

Decarbonization creates a tremendous production challenge. To "go green
fast," countries must produce an enormous amount of clean technology and
infrastructure in an incredibly tight time frame.[1] The numbers are nothing
short of astonishing. Renewable energy capacity must increase *fivefold by 2030*
if we are to stay on track to reach net zero by 2050.[2] This translates to roughly
$5 trillion in *annual* global investments in decarbonization, both to expand the
use of currently available technologies and to drive development of less mature
technologies. For comparison, in 2022 global investments reached a record
high of $1.3 trillion.[3]

The private sector alone cannot put up these numbers.[4] Government
investments—increasingly made through industrial policy—are critical. As
Karl Aiginger and Dani Rodrik note, "Industrial policy is back on the scene."[5]
And in particular, policymakers have come to see the benefits of *green* industrial policy—the "investments, incentives, regulations, and policy supports
designed to stimulate and facilitate the development of environmental technologies."[6] Green industrial policy is a critical tool in the climate policy

toolbox: it will help governments reorganize their economies for a decarbonized future.

Seen through the lens of existential politics, green industrial policy is not simply about creating green assets. It also has important *political* effects. Directly, it can generate new green asset owners who then become advocates for more ambitious climate policy. Indirectly, well-designed green industrial policy can create incentives for fossil asset owners to "convert"—substituting fossil assets for green assets where it is economically and technologically feasible to do so. Such conversion can set off cascading effects within industries and across jurisdictions—for instance, among those who work in green jobs or live in proximity to green assets. In other words, the success of "green transformations depends heavily on the *political effects of policies* aimed at bringing about transformation."[7]

Much of this work will happen at the domestic level through "behind the border" measures that incentivize the creation of green assets. These investments can help rebuild manufacturing capacity in countries where it has atrophied and promote economic competitiveness. However, such measures are not without risks. Political incentives to catalyze green asset creation can quickly shade into green protectionism—countries restricting imports or subsidizing exports under the guise of climate action. Though potentially politically useful, protectionism will ultimately slow the global diffusion of technology at precisely a time when we need to ramp it up.[8]

These are legitimate worries, but they are narrowly concerned about the direct effects of protectionism. Decarbonizing the economy will require undoing carbon lock-in, overcoming powerful obstructionists, and building *lasting* political coalitions that will support increasingly ambitious and costly climate policies. This transformation will require politicians to simultaneously play a long game and a short one. Domestic investments *behind* the border, rather than border tariffs, can contribute to this process.

Our collective timeline to avoid the worst effects of the climate crisis is exceedingly tight. We simply do not have time for every country to source and develop its own green technologies. As Nahm eloquently argues, countries and firms must find their "collaborative advantage"—their respective contributions to the global production of green assets that leverage existing comparative advantages.[9]

This chapter describes how to balance the need for enormous domestic investments—some of which may clash with global trade rules—with leveraging the efficiencies of the global supply chain. Governments will need more leeway to redirect capital to incentivize investments in green assets, thus expanding green asset owners as a political interest group within the existing trade regime. Unfortunately, there is neither time nor political will for a wholesale renegotiation of that regime.

So the question for existential politics is how to structure—and delimit—domestic investments in green assets. Policymakers will need domestic investments to generate broad political support,[10] but wholesale protectionism will slow the energy transition. Balancing the economic efficiency of global trade with the political benefits of behind-the-border measures and (some) protectionism is *the* trade-off that must be managed in existential politics. I call this "the Goldilocks challenge"—the middle path that veers neither too close to autarky nor too close to unfettered free trade at the cost of addressing the climate crisis.

Because green industrial policy will vary enormously by country, there is no single pathway for creating green assets, nor for meeting the Goldilocks challenge. In this chapter, I offer some guiding principles for catalyzing their creation in the face of obstructionism. Some readers may find this line of argumentation dissatisfying, since it remains at a general level. However, my argument demonstrates why existential politics is germane to understanding the realities of climate policy.

This chapter proceeds as follows. First, I explain why green industrial policy—defined as catalyzing the creation of green assets—is preferable to managing tons. Unlike managing tons, green industrial policy can provide immediate concentrated benefits to create new green asset owners and a supply of labor to build, maintain, and service those new assets. Costs are backloaded and diffused across the entire population, generating less opposition than carbon prices, whose costs are transparent and immediate. Investments can create positive spillovers, further driving down prices and promoting technology diffusion.[11] And the relative transparency of investments (dollars) compared to emissions (tons) makes accountability easier. I contrast these benefits with the European Union's carbon border adjustment mechanism, which *appears* to be focused on assets but is simply another form of managing tons, with its associated problems.

Second, I describe some of the ways in which green industrial policy can catalyze the creation of green assets and, importantly, the lessons from political science about how to overcome obstructionism from fossil asset owners.

Third, I discuss the Goldilocks challenge—meeting the twin challenges of investing in green industrial policy while using tariffs "responsibly." Economists tell us that protectionism is bad, since it hampers efficiency. But from a political perspective, tariffs have definite advantages: they can help grow industries domestically and avoid the unpleasant optics of government spending on foreign imports. Moreover, countries are already using tariffs, especially to reduce their reliance on China for renewables. In the vein of radical pragmatism, governments should consider how to deploy tariffs "responsibly" by maximizing political support for climate policy while minimizing efficiency losses.

Finally, I offer four principles that can inform the practice of global trade in the era of green industrial policy. Governments should deploy carrots, not sticks, in the short term. Incentives can help redirect capital, expand the number of green asset owners, and lessen political resistance to climate policy. In addition, policy should concentrate on decarbonizable industries such as electricity generation and auto manufacturing, which are ripe for conversion from fossil to green assets. Moreover, any good package of green industrial policy should develop the human capacity to provide services in support of the creation, installation, and maintenance of green assets. And finally, though it may sound counterintuitive, countries should use existing international rules rather than try to negotiate new ones. They should agree to certain practices without a wholesale reconfiguration of the World Trade Organization (WTO).

Brewing trade fights over green industrial policy clearly demonstrate that trade is the new frontier for climate policy. In an ideal world, countries would agree that maximizing economic efficiency is the pathway to deep and rapid decarbonization. But the realities of asset revaluation make such an approach infeasible, as it does not address the obstructionism of fossil asset owners.

What Are Green Assets?

Public discourse around climate policy often refers breezily to "green technology" and "green jobs." The most clear-cut category of green assets are the technologies that generate renewable energy and electrified transport: wind turbines and their component parts, solar photovoltaics, batteries, and the infrastructure needed to promote further electrification (grid expansion, charging stations).

In a completely decarbonized economy, all assets will be green. This chapter pays particular attention to sectors that can currently be "decarbonized": green technology is available and cost-competitive with fossil-based technologies.[12] Kupzok and Nahm define decarbonizable sectors as "industries

[with] . . . credible technological and economic pathways to decarbonizing their business models" as well as "an opportunity to gain competitive advantages through decarbonization."[13] Put simply, these are industries that *currently* have the capability to convert to nonpolluting products at a reasonable cost.[14] Frank Geels and his colleagues describe these as green technologies that are already in the diffusion phase where "radical innovations begin to spread when learning processes improve technical performance, lower costs, and generate clarity about how the new technology can align with consumer preferences and functional requirements."[15]

Power generation and automobile manufacturing are the most developed decarbonizable sectors; both technologies are available, largely cost-competitive, and rapidly diffusing.[16] There are other decarbonizable sectors, though they still need more investments to further incentivize their adoption. Kupzok and Nahm identify sixteen such sectors using three criteria: (1) sectors that use fossil fuel in production; (2) sectors that sell fossil fuel–based products; and (3) sectors that are energy-intensive. These include steel, cement, plastic, and rubber producers as well as energy-intensive products such as cars, heavy machinery, and agriculture.

In addition to providing the building blocks of the fossil-free economy, green assets also include climate-resilient assets, which are designed or retrofitted to cope with the effects of climate change. Climate-resilient infrastructure is "planned, designed, built and operated in a way that anticipates, prepares for, and adapts to changing climate conditions."[17] Bridges built at greater heights to account for sea-level rise, buildings that incorporate natural drainage systems, and permeable paving surfaces are all examples of climate-resilient assets.

Green assets are produced through global supply chains that can be roughly divided into four phases: extraction, processing of raw materials, component manufacturing, and assembly.

An electrified economy will require huge inputs of "critical minerals," such as nickel, lithium, copper, cobalt, and rare earth elements. The IEA estimates that if we are to reach net zero by 2050, demand for critical minerals will increase sixfold by 2040.[18] Each critical mineral faces distinct supply challenges related to quantity, geographic concentration, and costs.[19] While much of the political rhetoric has emphasized the need to diversify supply (the extraction phase), there are other challenges to ensure an adequate supply of inputs to meet soaring demand for green assets.

After extraction, critical minerals and other raw materials must be processed to create materials for green assets such as aluminum, steel, concrete, and polysilicon. Production capacity is unevenly distributed globally. For instance, China currently produces roughly 80 percent of global polysilicon and 70 percent of cobalt, while also leading in the global production of steel and cement.[20] Finally, critical minerals can be recycled post-use. As technologies for recycling improve and more inputs (for example, lithium-ion batteries) reach the end of their first life, demand for extraction may fall.[21]

These disparities have led policymakers to concentrate largely on extraction and production, emphasizing increasing extraction when possible, "made at home" goods and energy, and supply chain independence. For example, there are plans to develop new mines in northern Ontario's "Ring of Fire," which has extensive deposits of critical minerals. And Honda has recently agreed to invest $15 billion in Ontario to build four manufacturing plants to process the minerals needed for EVs, produce batteries, and assemble electric vehicles.[22] Ontario policymakers' vision is for a self-contained EV supply chain within the province that will includes extraction, production, and assembly.[23] These domestic investments build political support through the promise of jobs and investment.[24] They run the risk, however, of slowing production: mines will not be operational until 2030, but meeting this target will require the completion of access roads and other infrastructure.[25]

The creation and deployment of green assets will also require massive amounts of labor to manufacture, assemble, install ,and maintain these new technologies. The IEA estimates that clean energy manufacturing jobs would more than double if countries fully implemented their planned climate pledges.[26] Countries are already facing labor shortages for wind turbine and heat pump installers in Europe, the United States, and Australia.[27] Moreover, the energy industry requires more highly skilled workers than other sectors.[28] Thus, the need for labor is both broad (in terms of numbers) and deep (in terms of training).

Some of these workers may come from fossil asset industries, but not all will be able to migrate in this way. Indeed, much of the discussion about a "just transition" is about what to do about those who work for fossil asset owners. Existential politics will compel many to find new employment. In some cases, retraining may be possible, though relocation may not. But there are many other factors that make it difficult to simply swap out a fossil-based job for a greener one.[29] As I discuss later in the chapter, developing green industrial policies for services is a critical but underexamined component of existential politics.

Why Green Industrial Policy
Is Preferable to Managing Tons

Green industrial policy is both a functional response to market failures and a deliberate strategy by states to develop environmentally sustainable economic activities.[30] It includes policies such as subsidies, incentives, standards, regulations, public procurement, and direct investments as well as support for research and development.

Green industrial policy has several advantages over policies that manage tons. First, it builds political support by frontloading benefits and backloading costs. For example, the US Inflation Reduction Act provides $2,000 in subsidies to homeowners who wish to retrofit their homes with heat pumps. A recent study estimates that lower energy costs could reduce household energy costs by $73 to $370 per year by 2035, relative to the reference.[31] The Inflation Reduction Act's climate-related price tag is somewhere around half a trillion dollars, but these costs will change over time, as incentives will be paid out only when they are used.[32] (It is unclear how much of the Inflation Reduction Act will be repealed by President Trump.) By contrast, carbon pricing "highlights the short-term costs of climate action, jacking up the public's energy bills, while concealing the long-term benefits of addressing climate change—for the environment, public health, and the economy."[33]

Second, beyond the timing of benefits, green industrial policy can also distribute benefits to build further political support. Carbon pricing tends to concentrate costs and diffuse benefits.[34] Costs fall most heavily on large trade-exposed emitters—which is precisely why many governments agree to give them free allowances—while the benefits of reduced emissions are diffused across society.[35] In contrast, green industrial policy provides concentrated benefits in the form of government funding and support in its myriad forms (subsidies, incentives, and so on) while diffusing costs across all taxpayers.[36]

Third, green industrial policy can create positive spillovers. The clearest example of this logic comes from Chinese investments in renewables. China has spent decades investing in and developing specific capabilities in the solar and wind industries. Other countries (and the climate) have been the beneficiaries. Its sustained efforts helped radically reduce the production costs of solar photovoltaic cells[37] and wind turbines,[38] contributing to their falling costs worldwide. China currently manufactures roughly 80 percent of solar panel components (polysilicon, ingots, wafers, cells, and modules)[39] and 60 percent of wind turbine capacity.[40]

Their market dominance in these specific parts of the supply chain did not happen overnight. Other nations would have to invest similar resources and time to reach this level of market penetration—at a point in the climate crisis when time is in particularly short supply. And even then, gaining significant market share will be difficult in areas where China already dominates.

By contrast, spillovers in managing tons are more complex. The accounting headaches created by Scope 3 emissions are a type of (negative) spillover. Net zero pledges require a large multinational firm to either collect emissions data from its suppliers or fill in the blanks using off-the-shelf and often problematic data. In theory, the need for Scope 3 data could create an incentive for more supplier firms to measure and report emissions. In practice, it is difficult to ascertain the extent to which this has been the case.

Spillovers from carbon pricing and offsets are also often negative. They can create incentives to shift polluting activities to jurisdictions outside the regulated area (leakage). Moreover, within jurisdictions, the effects of other climate policies can adversely affect carbon pricing. For example, renewable energy generation requirements can drive down the demand for emissions allowances, and therefore the carbon price. And finally, these spillovers can be exacerbated when carbon pricing instruments from different jurisdictions are linked together.[41]

Finally, green industrial policy presents fewer accounting problems than managing tons. Dollars invested is a clear metric of effort. It is much more transparent than, say, trying to understand the size of a firm's Scope 3 pledge or the robustness of offset projects that governments use in their carbon pricing policies. To be sure, measuring dollars is not without problems. There are ongoing debates about whether developed country contributions to the UNFCCC's various funds are "new and additional" or simply the effect of shuffling funds. Money is fungible, so dollars cannot be a perfect measure of effort. However, dollars are both more politically legible and transparent than tons.

Consider the many funding mechanisms created by UNFCCC. While there have been peaks and valleys in the UNFCCC process, funding remains a clear and persistent failure. In 2009, fearing the impending collapse of the Kyoto process, developed countries promised "fast start financing" to fund developing countries' energy transition.[42] At the time, it was the most ambitious financial promise of the climate regime: $100 billion a year by 2020.[43] But developed countries missed the target, reaching $100 billion only in 2022.[44] In Paris, these same nations decided that $100 billion per year would be the floor in the run-up to the 2025 Conference of the Parties.[45] But in 2023 governments concluded that they were falling well short of this goal and that additional funds from developed countries were necessary.[46]

The transparency of these failures is critical: developing country governments and civil society can quickly and clearly demonstrate developed countries' lackluster efforts to provide funding. Insufficient funding undermines the legal principle of "common but differentiated responsibilities," which (nominally) obligates developed nations to take the lead in addressing the climate crisis, including funding developing countries' mitigation and adaptation efforts. Green industrial policy, as measured by government investments, affords a similar level of transparency, which in turn can facilitate holding governments to account. The adequacy or insufficiency of green industrial policy funding is much more legible than evaluating whether policies that manage tons are robust.

Carbon Border Tariffs: Another Form of Managing Tons

Recall the discussion of the European Union's carbon border adjustment mechanism in chapter 3. The CBAM is a trade-oriented form of carbon pricing. It imposes a fee on the embedded carbon of energy-intensive imports from countries without a carbon price (or with a lower price). In its current pilot phase, importers are required to measure and report the embedded carbon of purchased products but do not actually have to purchase allowances. The financial component—when importers must pay out—does not begin until 2026.

The CBAM is the latest version of managing tons, and unsurprisingly, it suffers from the same weaknesses: political controversies and big implementation challenges. Politically, the CBAM has already kicked up an international fuss. India and China have raised objections to the policy, and continued diplomatic discussions between the European Union and India have yet to produce a mutually acceptable outcome.[47] The Indian government has plainly stated its concerns about protectionism: "Any unilateral measures taken to combat climate change should not constitute a means of arbitrary or unjustifiable discrimination or disguised restriction on global trade."[48] Developing countries have also argued that CBAMs are fundamentally unjust because they punish developing country economies for carbon-intensive production rather than funding a green transition.

In terms of implementation, preliminary research indicates that, like other carbon pricing policies, the CBAM will do little to reduce emissions. The

Asian Development Bank estimates that emissions will be only 0.2 percent lower than they would by simply using emissions trading *without* the CBAM.[49] At the same time, it will have a disproportionate impact on middle- and low-income countries.[50] This reproduces the political backlash conditions associated with fetishizing tons—transparent, up-front costs with little predicted effect on emissions. Political pain for minimal climate gain is not a recipe for ambitious climate policy.

And of course, like other policies that manage tons, the CBAM requires accurate measurement of the GHGs associated with the production of each imported good. As chapters 3 through 5 demonstrate, measuring emissions creates many challenges. Exporters must either measure and verify the emissions generated at the site of production or calculate emissions based on sectoral benchmarks, which estimate emissions based on the averages associated with each technology.

It's also worth noting that the US-EU "green steel" agreement currently appears to be a casualty of the CBAM. In 2021, the United States and the European Union began crafting the Global Arrangement on Sustainable Steel and Aluminum—essentially, a club of green steel producers. Green steel is primarily produced with electric arc furnaces powered by renewable energy rather than coal-fired blast furnaces.[51] The US-EU agreement aims to lower tariffs on steel and aluminum imports between the trading partners, but not for other countries where production methods are more carbon-intensive.

The agreement is a test run for much touted "climate clubs": beneficial trading arrangements among governments with similar climate preferences, contingent on certain minimum climate policies or goals.[52] But the green steel agreement appears to be stuck on the thorny issue of measurement. The United States has proposed defining "dirty" steel as that which exceeds the carbon intensity level of the highest emitters in the United States. The European Union, by contrast, is proposing rules consistent with the CBAM, which would require calculations of embodied carbon.[53] These two approaches, which will ultimately determine who pays tariffs and at what levels, are incompatible and have contributed significantly to stalled negotiations.[54]

Despite these problems, CBAMs seem poised to become the next new thing in climate policy. Australia is in the middle of a review and consultation process on carbon leakage—the main problem that CBAMs seek to correct.[55] Japan, Canada, and the United Kingdom are also considering some form of carbon border tariff.[56] And there have been legislative proposals for border tariffs in the United States as well.[57]

What Does Catalyzing Green Assets Look Like in Practice?

Green industrial policy can be many different things. Subsidies and local content requirements tend to get a lot of attention, since these often violate global trade rules and end up before the WTO Appellate Body. But it is important to note that there are many types of green industrial policy that, generally speaking, do not conflict with WTO rules.

For example, governments can use procurement policy to drive demand for green assets. The World Bank estimates that governments collectively spend $US9.5 trillion per year on procurement.[58] In OECD countries, almost 15 percent of GDP goes toward procurement.[59] Climate-friendly procurement can take many forms—ranging from electricity purchases to furniture and cloud computing. Many governments have committed to converting government-owned vehicles from internal combustion to electric. The 2019 EU clean vehicle directive sets minimum procurement targets for the share of clean vehicles through 2030. Similarly, Canada has set a goal of 100 percent zero-emissions vehicles by 2030. These policies can help boost demand for green assets.

Governments can also use more "disruptive" types of green industrial policy to facilitate the phaseout of fossil assets.[60] They can levy taxes, reduce or eliminate incentives, or mandate phaseouts. The European Union has committed to phasing out internal combustion engines by 2035.[61] Although there is pressure to roll back this commitment, most automakers are backing the policy, noting that it provides "a solid basis for long-term competitiveness and investment into the region."[62]

Subsidies for consumers and services are also permitted under the trade regime. The former can stimulate demand for green assets and are WTO-aligned since they do not distinguish between domestic and foreign producers. The latter is critical for expanding the human resources required for decarbonization (discussed later in the chapter).[63] Governments could therefore subsidize the installation of renewable energy or training for other activities necessary for decarbonization. They could invest in expanding employment in the "care economy"—sectors like health, education, and elder care, which are both in demand and sources of low-carbon jobs.[64]

But of course, in the face of fossil owners' obstructionism, implementing green industrial policy is easier said than done, even when it does not violate trade rules. Though countries will face distinct challenges with obstructionism, political science research offers some important insights into effective strategies for targeting decarbonizable industries.

First, the size of decarbonizable industries matters. Kupzok and Nahm analyzed climate spending in OECD countries and found that governments' climate investments "can be explained by the relative size and influence of . . . the decarbonizable sector."[65] Utilities, the automotive sector, and other energy-intensive manufacturing industries can exert pressure on governments to invest in helping them convert fossil assets to green ones. Admittedly, this presents a bit of a chicken-and-egg problem: bigger industries lead to more investment, but investment builds the size of the industry. Nonetheless, this suggests a mechanism for positive feedback: as certain types of green asset owners expand, so too will government investment in them.

Second, intrasectoral competition can accelerate the implementation of green industrial policy. And the converse is also true: collaboration can slow progress. Interfirm competition can be an important source of ambitious policy. Meckling and Nahm show that state intervention to promote electric vehicles was more effective in the United States than in Germany precisely because there was less coordination among firms within the US automotive industry. This lack of coordination among incumbents prevented unified opposition to increased fuel efficiency standards and emissions.[66]

Conversely, collaboration among firms within a given sector allows them to present a united front with governments, slowing progress. Charles Bain shows that, because national institutions facilitated firm-state interactions, the cleanest utility in Germany, E.ON, aligned its climate stance with that of RWE, one of the dirtiest utilities.[67] These institutions of "concertation" incentivized utilities to present a united front on climate policy, even though they faced very different costs for increasingly stringent regulations.

Finally, there are ways to quiet opposition from fossil asset owners. While it may seem objectionable to some, compensation can help diffuse opposition. In the German coal phaseout, coal workers, utilities, and other energy-intensive industries received €40 billion in compensation, which garnered the acquiescence of previously opposed labor unions.[68] In 2018, the Spanish government, led by the Socialist Party, successfully negotiated the closure of twenty-eight coal mines, contingent on providing €250 million to support the affected regions over the following decade.[69] There is additional evidence that compensation to voters in areas adversely affected by climate policy, such as coal-mining regions, garners political support.[70]

Global Trade in an Era of Green Industrial Policy

The brewing global trade war around EVs is a prime example of how trade is the new frontier for climate policy. In late 2024, the United States and Canada placed 100 percent tariffs on EVs imported from China, citing China's "unfair trade practices."[71] The European Union then followed suit, instituting a 45 percent tariff, which has split member states. China responded by raising tariffs on European brandy[72] and pork[73] and opening an anti-dumping investigation into Canadian canola oil.

The EV controversy shows the contours of the Goldilocks challenge. China is far and away the largest global producer of EVs, and it produces lower-cost models, which are less abundant globally, particularly in the United States and Canadian markets. Private cars and vans generate about 10 percent of global emissions; the IEA estimates that their emissions must fall by 6 percent per year by 2030 to stay on track with the target of net zero by 2050.[74] If countries truly want to decarbonize and meet their 2050 net zero targets, selling cost-competitive Chinese EVs seems like a no-brainer.

But there are political benefits to enacting these tariffs that seem to overshadow concerns about complying with WTO rules for the United States and Canada.[75] First, the US and Canadian tariffs were rolled out as both countries were nearing elections where the incumbent party's power was under threat. The Democrats subsequently lost the 2024 election to Donald Trump, and there is speculation that the Liberals in Canada will lose power in the next election, which will be held by October 2025. At the time of writing, President Trump has threatened to enact or already enacted major tariffs on imports from Canada, China, and Mexico, and more tariffs are likely on the horizon.

Second, cheap Chinese imports threaten domestic competitiveness. For this reason, the US auto industry supported the EV tariffs.[76] The European Union, though more divided, also cited concerns about unfair advantages when enacting tariffs.[77]

The waning power of the WTO, coupled with the rise of green industrial policy, provides an opportunity for creating new practices that meet the Goldilocks challenge—allowing for some protection to facilitate political coalition building among green asset owners—without rewriting the rules of the global trade regime.

Domestically, green industrial policy will vary widely based on current economic activities, resource endowments, state capacity, and interest group politics, among many other factors. But guiding principles informed by existential politics can help meet the Goldilocks challenge, promoting not only the production of green assets but also the forming of political coalitions to build support for more ambitious decarbonization policies.

Behind the Border Mechanisms: Carrots, Not Sticks, in the Short Term

Existential politics requires engaging with the political reality that green assets must be built in the face of substantial obstructionism. In some cases, this will mean that governments need to put their thumbs on the scale by providing incentives that may be in tension with international trade law. In a recent paper, colleagues from the climate modeling community and I find that the implementation of near-term "carrots"—operationalized as the current subsidies for clean energy in US federal policy—can accelerate emissions reductions in comparison to a policy approach that uses only "sticks." Not only does this approach incentivize action, but it can also lower future decarbonization costs by catalyzing technological change and uptake.[78]

These "behind-the-border" measures should seek to maximize near-term political benefits while minimizing efficiency losses. Governments should build on existing capabilities rather than on developing all parts of the supply chain. This is the "collaborative advantage" that Nahm describes: Chinese solar PV firms collaborated with American R&D firms to incorporate new materials into production processes and with German firms as a source for components.[79] Each country built on its comparative advantages and drew from others as needed.

Policies should seek to align economic and production goals. For example, green industrial policy may seek to develop new domestic supply chains, or it may focus instead on fostering competition among existing firms so that they can integrate into global supply chains.[80] The former strategy is likely to require more protective measures, since fledgling industries struggle to compete with more mature ones. The latter can, counterintuitively, incentivize opening markets to ensure that domestic producers can compete with foreign competition. For example, as China's EV market matured, its "dual credit" system for EVs complemented targeted subsidies with a credit system open to all manufacturers selling in China. Virtually all credit recipients were Chinese firms, except Tesla. This opening spurred Chinese firms to compete with one of the leading EV manufacturers, further developing the domestic market.[81]

Technological development matters too. Policies to develop new or nascent technologies (for example, direct air capture) will differ from those that seek to scale relatively mature ones (such as solar or wind energy). With mature technologies, uncertainty is no longer centered on technological capabilities but rather on scaling and entrenching those technologies.[82] For instance, renewable electricity is now easily generated with available technologies; the remaining challenge is to expand and modernize to promote rapid decarbonization.

Incentives are critical to restructure investment flows, but sticks will also be needed in the medium term. Climate models of emissions in the United States show that "relying solely on carrots without shifting to sticks is unlikely to substantially bend the emissions curve and achieve deep decarbonization by mid-century."[83] As discussed earlier, sticks can include taxes, fees, mandated phaseouts, removal of incentives, and technology or performance standards.

Focus on Converting "Decarbonizable" Industries

Decarbonizable industries—like electricity generation and auto manufacturing—are ripe for conversion, which occurs when fossil asset owners transform their fossil assets into green ones.[84] Individual firms' investments in green assets can drive competitors to follow suit, setting off a cascading effect within an industry.[85] In other words, intrafirm competition matters for further creation of green assets. If firms within an industry face different costs for regulation, those facing lower costs may prefer more stringent regulations, which give them a competitive advantage.[86] This dynamic can help create a positive feedback cycle of conversion.

We need rapid, near-term reductions, so focusing on decarbonizable industries is critical. A recent study estimates that given new "adverse information" about decarbonization—including a reduced carbon budget and slower-than-expected deployment of renewables—emissions reductions should be closer to 3 gigatons of carbon ($GtCO_2$) per year until 2035, which is seven times the reference scenario rate of 1.5 degrees Celsius.[87] The sooner these reductions begin, the less costly they will be. Ramping up the diffusion of fossil-free technologies can happen much more quickly than the development or scaling of new ones.

Moreover, political science research demonstrates that the diffusion phase of new technologies is a mobilization moment for opponents.[88] As new technologies' cost decreases and deployment grows, opponents start to feel threatened and push back. Decarbonizable industries are often engaged in this fraught moment of backlash. For example, as Hanna Breetz and her colleagues document, the main electric utility in Arizona backed the policy of net metering—until its use became widespread in the state. Net metering allows customers with rooftop solar to sell electricity back to the grid, effectively undermining the utility's monopoly over power generation. Initial incentives prompted a rapid expansion of net metering—customers grew almost twentyfold in four short years. This growth prompted the utility to roll back its financial support. It withdrew incentives and imposed new fees, ultimately raising the relative cost of solar energy in the state.[89] At these critical junctures, concerted efforts through domestic green industrial policy can prevent backsliding or incremental changes that fail to yield full-scale technological shifts.[90]

There is no one-size-fits-all policy that can target all decarbonizable industries.[91] Policy choice will vary by country depending on a variety of factors. However, the final two principles consider specific types of policies that can help target decarbonizable industries.

Industrial Policy for Services

We tend to think about industrial policy as manufacturing and production. However, it also includes services, which will be a critical component of the energy transition. As Réka Juhász and her colleagues note, "It is almost a statistical certainty that the bulk of . . . jobs [created by industrial policy] will have to be generated in services."[92] Indeed, the International Energy Agency estimates that for the development of a 50-Megawatt (mW) solar PV plant, 56 percent of person-days of labor will focus on operation and maintenance.[93]

Indeed, creating a decarbonized economy will require a huge workforce.[94] Charging stations need to be built, and electricity grids must be expanded. Solar PV installations and turbines will require maintenance and upkeep. Buildings will need retrofits both to get off natural gas and to enhance energy efficiency. Existing infrastructure will need upgrades to improve resilience.

In an analysis of green industrial policy in the United States, Todd Tucker declares that "any industrial policy regime worth its salt would spend at least half of its energies on the service sector."[95] The service sector employs almost

80 percent of working Americans, while manufacturing represents only 12 percent of US jobs.[96] This is the case elsewhere as well. In OECD countries, an average of 70 percent of working people are in the service sector.[97] Rodrik finds that a similar pattern holds for other advanced and emerging economies, such as China, Japan, Mexico, and Turkey.[98]

Put simply, green industrial policy must move beyond measures to stimulate manufacturing to include policies to stimulate services.[99] Yet, despite calls for a just transition, we have little in the way of concrete models of what industrial policy for services could look like, much less evidence about which policies work.

Rodrik offers a proposal to put subnational policymakers in the driver's seat. His proposal assumes a high level of uncertainty and information asymmetry between levels of government. Thus, one of the main challenges of service-oriented industrial policy is to figure out the types of jobs that are "best suited" (however policymakers choose to define this) to a given jurisdiction. He cites the American Rescue Plan "challenges" as a potential model: federal funds were allocated for local and regional good jobs challenges in which local groups submitted bids to be funded in order to act as "organizational hubs for federal-supported initiatives."[100]

Use Existing Tools First

There is a long-standing and large body of literature examining the tensions between the trade regime and climate protection.[101] A slew of proposals offer ways to reconcile these tensions, such as designing principles for a WTO-consistent border adjustment tariff, reinstating provisions in the WTO Agreement on Subsidies and Countervailing Measures to allow for green subsidies, and even issuing a blanket waiver across all multilateral trade agreements to accommodate the exceptional circumstances of the climate crisis.[102] There is even a new agreement on climate change, trade, and sustainability that has been signed by New Zealand, Switzerland, Norway, Costa Rica, and Iceland, which have agreed to eliminate tariffs on over three hundred environmental goods.[103]

Policymakers and advocates should resist, for now, the urge to create more multilateral rules to govern trade and climate or reopen negotiations on existing agreements, for several reasons. First and most importantly, negotiating multilateral agreements—either creating new agreements or reforming existing ones—is an incredibly time-consuming process. This is time that we collectively do not have. Second, depending on the actors involved and the effectiveness of fossil asset obstructionism, any agreement is likely to be watered down, in line with the preferences of the least ambitious governments.[104] Third, negotiating new agreements can provide rationales for delay. States can claim that they can accelerate behind-the border-measures only once the necessary reforms are agreed upon.

Seen through a different lens, trade conflicts over green industrial policy can create "productive tensions" that prompt governments to sort out their differences through diplomatic channels. Of course, there is the danger that protective measures will set off a negative feedback cycle of beggar-thy-neighbor policies. But the economic interdependence of advanced economies creates counterpressures to resolve disputes.

There is useful precedent for this approach. In 2012, the European Union decided to include aviation emissions in its emissions trading system. Initially, this would have required all airlines to purchase allowances to cover emissions for flights originating or terminating in the European Union, even if the firm was located outside EU nations that are covered by the ETS. The United States, Russia, China, India, and Saudi Arabia instructed their carriers not to comply with the EU law.[105] A global trade war seemed very likely. Seeing the damage that this would inflict, the European Union quickly agreed to reduce coverage to intra-EU flights only, while insisting, in exchange, on the negotiation of a global agreement to regulate airline emissions. Four years later, countries adopted the Carbon Offsetting and Reduction Scheme for International Aviation (CORSIA) agreement. Quite simply, the European Union was not willing to blow up the global trade regime in exchange for regulating aviation emissions.

Governments continue to sort through the political challenges that protectionism creates. After initial concerns about generous subsidies and local content requirements in the US Inflation Reduction Act, the European Union has now entered into talks with the United States to ensure that European producers are not unfairly disadvantaged. It has also enacted the Net-Zero Industry Act to boost EU production of green assets.[106] South Korea is pursuing a similar diplomatic route.[107]

The US-EU negotiations are consistent with Timothy Meyer and Todd Tucker's proposal that countries simply give each other more leeway in making domestic climate policy. Specifically, they suggest that governments "exercise forbearance in the decision to bring WTO challenges" and, more generally, grant "more deference to national regulators for climate-related measures."[108]

Trade Is the New Frontier of Climate Policy

In 2014, author and activist Naomi Klein described trade and climate regimes as "two solitudes." Even though they are deeply interconnected, the set of international rules governing each regime was created largely independently of the other.[109] There have been long-standing tensions between trade and climate policy, but the turn toward green industrial policy has sharpened the need to reconcile these two solitudes.[110]

The trade regime—with its ability to supercharge the production of green assets—has become the new frontier for climate policy. The test will be to meet the Goldilocks challenge: to balance the benefits of domestic carrots with the need to leverage all the production efficiencies that global trade creates. An emphasis on services is clearly needed, both to accelerate the energy transition and to provide the political benefits that asset revaluation demands. But this is largely uncharted territory that governments must continue to manage through self-restraint and diplomatic negotiations. Doing so will require a new kind of international cooperation—cooperation focused on assets, not tons.

8

The Future of Global Climate Policy

IF THERE IS ONE MESSAGE to glean from this book, it is that assets, not tons, must be the mainstay of climate policy going forward. Managing tons reduces climate policy to technocratic questions about measurement, verification, and commodification. It is a complex endeavor, often full of educated guesses. The technical nature of managing tons makes it the province of a small group of experts, and therefore particularly vulnerable to massaging the numbers.

The measurement challenges are just the first problem. There is little to no regulation of the voluntary market or of corporate net zero claims. Thus, it is no surprise that both policies are rife with greenwashing and sometimes outright fraud. While carbon pricing is regulated by governments at the regional, national, and international levels, it has had limited effects on emissions, and even fewer on decarbonization.

Instead of continuing down the problematic and not terribly effective path of managing tons, I have suggested a different way forward—regulating assets and engaging directly with the political power of their owners. Fossil asset owners are the Goliath in the climate policy story. They are precisely what makes climate policy such a tough nut to crack—and why radical pragmatism is a useful starting point.

The solutions I have proposed are radical in that they get to the root of the problem—the political and economic conflicts between fossil and green asset owners. But they are also pragmatic, since they do not require a wholesale reinvention of global institutions. Given the breadth of the obstructionism of fossil asset owners and the incredible time pressures we face, a wholesale remaking of the global order is not a possibility. Instead, we must work in the short term within the constraints of the current system, even as we acknowledge its many flaws and limitations. The climate crisis requires urgent and drastic action, so radical pragmatism, while by no means ideal, will have to suffice for now.

Next Steps: Climate Policy Focused on Assets

As I finish writing this book, the global climate regime is under serious threat. Donald Trump has just been inaugurated for a second term as US president. He will surely gut domestic climate policy and has already withdrawn from the Paris Agreement, as he did in his first term. The UNFCCC process is facing a legitimacy crisis. More broadly, the failures of neoliberalism have contributed to growing inequality, war, and political polarization—all of which exacerbate climate change, making the need for progress all the more urgent.[1]

The grim realities of the present day provide a political moment for implementing new asset-focused climate policies. Indeed, radical pragmatism could strike a fortuitous balance, being palatable to politicians and technocrats comfortable with the familiarities of incrementalism, on the one hand, and rhetorically appealing to regular people in search of material improvements to their lives, on the other. In an era of rising populism, messages like "tax the rich" and "boost domestic production" should resonate. There are several measures that governments can take in the short term to reorient climate policy toward assets.

First, governments can leverage the political momentum around taxation. There are already windfall taxes on fossil fuel companies in six European nations, and countries are moving forward with a minimum corporate tax, as detailed in chapter 6.[2] The drumbeat for greater taxation continues, both through the G20 and the new UN Tax Treaty. In 2024, the G20 agreed—in principle—to cooperate on taxing the ultra-rich, though whether and how this will actually be implemented is unclear.[3] The UN Tax Treaty is still in its preliminary phases, but has identified "addressing tax evasion and avoidance by high net-worth individuals" as part of its remit.[4]

Some of this rhetoric is likely to wane with the Trump presidency, given that Congress enacted historic tax cuts to the wealthy during his first term. But the cost-of-living crisis seems unlikely to abate, and with it, the call for taxing wealthy individuals and corporations will continue to surface as a solution.

Second, countries can take unilateral steps to dismantle ISDS protections or reform the terms of current investment treaties. At this stage of the climate crisis, investment protections for fossil asset owners are sheer madness. But as with all policies that threaten fossil asset owners, the politics are not so simple. There are signs of progress, such as the European Union's withdrawal from the Energy Charter Treaty and the exclusion of the ISDS between the United States and Canada in the USMCA. But there is more that countries can do.

Countries can exit these treaties or engage in negotiations to reform them. Because many ISDS protections are extended through *bilateral* treaties, the ability to withdraw from or reform these treaties is relatively straightforward, since there are simply fewer parties to deal with and one party can withdraw unilaterally. Reform may require agreement between only two parties.

Both withdrawal and reform are happening with increasing frequency.[5] Roughly 10 percent of all international investment agreements have been re-negotiated, and 2 percent have been terminated altogether.[6] Research indicates that the more frequently countries are subject to claims by foreign investors, the more likely they are to renegotiate agreements. For instance, between 2011 and 2014, Indonesia was the target of several large claims brought by firms in the United Kingdom, Saudi Arabia, and the Netherlands. It subsequently an-nounced its intention to withdraw from all sixty-seven bilateral investment treaties it had signed; to date, Indonesia has terminated twenty-four treaties and renegotiated two.[7]

States can also withdraw from the International Center for Settlement of Investment Disputes (ICSID), which serves as the institutional forum for ISDS arbitration; Bolivia, Honduras, and Venezuela have already done so.[8] These actions may be heavy political lifts domestically, but they are attractive since they do not require cooperation on a global scale. Moreover, every with-drawal or reform helps erode the incumbency of fossil asset owners in the international investment regime.

Third, advanced economies can lower the temperature on the trade ten-sions with China by concentrating on domestic investments, particularly in services. Catering to domestic interests can be accomplished through tariffs, which will raise consumer costs, or through investments in cultivating domes-tic services. Chapter 7 makes clear that tariffs on green assets will only slow the energy transition. However, investments in the shrinking middle class through job training can make climate policy more palatable.[9] This in-cludes training labor to service the growing stock of green assets as well as expanding the care economy (health care, child and elder care, education). For instance, if we are going to "electrify everything" to accelerate decarbon-ization, we will need *many* more electricians.[10] To date, industrial policy has been almost exclusively focused on "making stuff." But given that, on average,

70 percent of employment in OECD countries is in the service sector, training must be addressed.[11]

Managing Tons: Reducing Harm in the Short Term

If climate policy turns toward assets, governments are left with the tricky question of what to do about the lackluster policies that manage tons. Net zero pledges, offsets, and carbon pricing aren't going anywhere. Repealing these policies would be difficult and frankly a waste of finite political resources. International carbon markets are inscribed in Article 6 of the Paris Agreement. Axing them would require renegotiating the agreement, which simply won't happen. In addition, there are now seventy-five carbon pricing laws in effect at the regional, national, and subnational levels, covering roughly one-quarter of global emissions.[12] Widespread repeal of these laws is hard to imagine. Similarly, it is unlikely that governments or firms will roll back their net zero pledges, particularly if they are cheap talk and provide good PR at relatively little cost.

In the short term, the best approach to managing tons is harm reduction—reducing the negative consequences of these policies without engaging in the unrealistic expectation that they will be reversed. Cullenward and Victor call for a "right-sizing" of markets to reduce their prominence in the portfolio of policies that address climate change.[13] All policies that manage tons should be right-sized, both in our expectations about their capacity to reduce emissions and, critically, in the extent to which they are perceived as solutions. To this end, there are several measures that governments can implement.

Harm reduction for carbon credits is very tricky. As Cullenward and Victor note, "The problems with offsets are structural, not experiential, and therefore offsets have limited potential for reform."[14] I have argued that it's time to get rid of all nonpermanent offsets.[15] But what to do about Article 6.4, the Paris Agreement Crediting Mechanism (PACM)?

There are several measures that can try to limit the damage of this new market. First, the Supervisory Body should limit the types of projects allowed in the PACM. For instance, renewable energy, energy efficiency, and fuel-switching activities should be business as usual in a decarbonizing world. Unless they demonstrate ambitious improvements over the status quo *and* are aligned with the Paris temperature goal, these project types should not be permitted. It appears that the rules are moving in this direction, but further detail from the Supervisory Body is needed to ensure that this is the case.[16]

Forestry projects, which are at risk of reversal and systematically overestimate reductions, should also be excluded.

Second, the Article 6 rulebook has already allowed for grandfathering in some CDM credits, including extremely dubious credits from afforestation and deforestation projects.[17] But these could be devalued. For example, valuing each CDM credit at half an Article 6.4 credit would disincentivize their continued use. The sooner these very questionable credits are out of circulation the better.

Third, governments should reduce the number of carbon credits that regulated entities can use to reach their targets. The schedule for reduction should be accelerated, reaching zero as quickly as possible. There is precedent for this: the EU-ETS has not allowed offsets since Phase 4 began in 2021. There will be pushback from fossil asset owners, project developers, and others involved in carbon markets, but since offsets are often a small part of the overall reduction, this proposal is not out of the realm of the possible.[18] The reduced demand would have the added effect of diminishing the profits of those in the carbon market industry and, potentially, some of their political power.

Fourth, for all the reasons detailed in chapter 4, I do not think that the voluntary market can be salvaged. Harm reduction requires government regulation. Governments can regulate what types of claims can be made and how these commodities are structured and sold. For example, in late 2024, Zimbabwe introduced a bill that would create a carbon trading unit in its Climate Change Management Department. The unit would be responsible for registering and monitoring offset projects from the voluntary carbon market and ensuring their environmental integrity.[19] This puts the onus on sellers to demonstrate the veracity of their claims to regulators rather than on buyers to exercise due diligence.

In creating oversight, governments must ensure that these new regulators are *entirely independent* from extant organizations in order to resist the temptations of the revolving door. As I document in chapter 4, the melding of voluntary and compliance markets over time has made independent regulators particularly important.[20] Similarly, governments that create their own offset programs should not adopt methodologies from the voluntary market. This would simply import bad rules into new programs.[21]

Fifth, emissions trading can also be reformed to reduce speculation and gaming. Clearly, governments should stop giving out free or reduced-cost allowances. But like abolishing fossil fuel subsidies, this is a policy no-brainer with a difficult political path. Even the European Union—the global leader in emissions trading—still distributes some free allowances.[22] A central carbon bank, like the European Union's Market Stability Reserve, could also help maintain the value of allowances and, critically, guard against oversupply. Again, however, this is a difficult political lift.

A more feasible reform would be to make emissions trading schemes function more like taxes. Governments can create price collars to ensure a minimum price, while providing industry with assurances that costs will not skyrocket. They can prohibit the use of offsets and the banking and borrowing of allowances. These actions would require regulated entities to true up their accounts each year—effectively subjecting them to a tax.

Finally, governments should refrain from linking existing carbon markets.[23] This idea came to the fore in the mid-2010s as the Kyoto Protocol was losing momentum.[24] By linking national and subnational markets, governments could theoretically circumvent the messy politics of the UNFCCC. But linking markets is a risky idea. It *could* work, but only once individual markets are well established. As chapter 3 demonstrates, well-functioning carbon markets require the elusive trifecta of political will, extensive bureaucratic capacity, and a centralized body to guard against oversupply. Currently, this trifecta exists only in the European Union. Not even California—a global leader in climate policy—has been able to tame the problem of oversupply. Linking carbon markets is much more likely to spread the problems and weaknesses of individual programs across jurisdictions than it is to create a global price on carbon.

In the net zero arena, harm reduction requires government regulation to crack down on corporate greenwashing. For example, Canada passed Bill C-59 in 2024, amending the Competition Act to target greenwashing by including provisions about truth in advertising.[25] Oil and gas companies strenuously objected.[26] Once the bill passed, the Oil Sands Pathways Alliance—the group of Canadian fossil fuel companies pledging to go net zero—scrubbed their entire website, presumably for fear of running afoul of the new rules.[27] Canada is one of twenty-nine countries that now have some form of regulation on net zero claims that covers approximately 40 percent of global GDP.[28] Governments should expand the breadth and stringency of these rules.

Beyond truth in advertising, governments can also create rules governing disclosure about climate risks and pathways to net zero. They can also use net zero standards as a condition for procurement and lending.[29] For example, the European Investment Bank—the lending arm of the European Union—now requires all financing to be Paris-aligned.[30] This means that loan recipients must demonstrate how they will reduce their emissions.

Regulation is an important avenue for harm reduction, but as Derik Broek-hoff and Cleo Verkuijl warn, "holding actors accountable . . . should not be an end in itself. Such regulation risks becoming its own kind of distraction, as governments expend resources to police . . . commitments that could be better spent directly regulating emissions and implementing broader decarbonization programs."[31] In other words, harm reduction can only get us so far.

The Future of the UNFCCC

The Paris Agreement is teetering on the brink of complete irrelevance. Governments have been negotiating what to do about climate change for three decades. In the meantime, the climate crisis has arrived—in full force. The Paris Agreement may have been a diplomatic success, but it has been a climate failure. Countries' NDCs will reduce emissions only 2.6 percent below 2019 levels by 2030.[32] The UNFCCC Secretariat, uncharacteristically forceful in its language, notes that "greenhouse gas pollution at these levels will guarantee a human and economic trainwreck for every country, without exception."[33]

The COP process is similarly problematic. Despite its dismal track record on emissions, the COP keeps growing. Each year, it creates new work streams, commitments, funds, and governing bodies, as well as an ever-changing repertoire of ad hoc committees. And the annual COP meeting consumes enormous amounts of human and economic resources and attracts international attention and media coverage. But of course, expansion is not evidence of success. Indeed, the opposite is true. As it becomes more labyrinthine, the COP appears to be collapsing under its own weight. The slavish commitment to process fails to produce meaningful outcomes, further undermining the legitimacy and relevance of the COP and the UNFCCC as a whole.

COP29 is just the latest example. Even before its disappointing outcomes, it was already exhibiting signs of distress. COP29 was held in a petro-state run by a strongman (as was COP28, held in the United Arab Emirates). Many major economies—including the European Union, a traditional climate leader—did not send their head of state. Even the small island nation of Papua New Guinea, which is extremely vulnerable to sea-level rise, boycotted COP29, stating that attending would be "a total waste of time."[34]

The major outcome of the conference was a paltry financial commitment from the developed world. After two weeks of chaotic negotiations, countries agreed—thirty-three hours past the deadline—to provide "at least $300 billion per year by 2035 to developing country Parties for climate action."[35] This figure pales in comparison to the $1.3 trillion that developing countries asked for, and the $2.3 trillion to $2.5 trillion by 2030 that the Independent High-Level Expert Group on Climate Finance estimates is necessary to meet the Paris Agreement target of net zero by 2050.[36]

These failures have prompted calls for reform. In an open letter to the UNFCCC, a group of academics, advocates, and policymakers offer seven proposals, including requiring host countries to demonstrate their commitment to phasing out fossil fuels, shifting to smaller and more frequent meetings dedicated to implementation, and improving transparency and accountability mechanisms.[37] If implemented, which is unlikely, these reforms could make some marginal improvements.

Existential politics explains why these reforms will have little effect and, more broadly, why the UNFCCC will not be the engine of decarbonization: collective action on managing tons does not address the conflicts created by asset revaluation. If states are to pursue a new international strategy for climate focused on assets rather than tons, then they will need to shift the locus of climate policy away from the UNFCCC. How can we "right-size" this institution so that it is no longer the focal point for global climate rulemaking?

The first step is to treat it like a fire burning out of control by depriving it of oxygen. The unnecessary fanfare of the COP cannot be overstated. Every year tens of thousands of people attend in order to negotiate, network, and advocate. COP28, the last year for which data are available, drew roughly seventy thousand participants, including governments, international organizations, NGOs, and the media.[38] This annual ritual feeds the belief that the UNFCCC is "the only game in town." It does not have to be.

Instead of seeing the UNFCCC as *the* locus for climate policy, its role should be more circumscribed. It should remain the institution to which countries report on their national efforts. If history is any indication, it is extremely unlikely that this transparency will generate meaningful efforts to name and shame laggards—often a key element of many multilateral institutions.[39] But the UNFCCC can continue to gather the raw data about governments' collective efforts and about the work that remains.

The UNFCCC can also continue to provide ancillary funding for mitigation and adaptation efforts for the developing world. It currently has six financial mechanisms.[40] The largest is the Green Climate Fund, which approved $52 billion in projects between 2015 and 2023 and disbursed $4 billion.[41] The Special Climate Change Fund, which was created in 2001 to assist countries with adaptation, has disbursed just $400 million in grants.[42] Of course, developing nations need (and indeed are owed) all the money they can wring out of the developed world, but again, we should not expect these modest sums to do any heavy lifting when it comes to catalyzing the creation of green assets.

The UNFCCC should also remain the primary venue for discussions around adaptation. In 2010, countries decided to create and submit national adaptation plans to the UNFCCC. Five years later, the Paris Agreement began a process to establish a global goal on adaptation to help countries increase their adaptive capacity, decrease their vulnerability, and enhance their resilience.[43] While adaptation is, by definition, location-specific, the UNFCCC can continue to serve as the venue where governments share information and plans.

In the end, the major changes around the UNFCCC will be twofold. First, we must adjust expectations about what this multilateral process can achieve vis-à-vis decarbonization. And second, we must shift the mental model away from managing tons and toward creating green asset owners and constraining fossil asset owners.

Calling for a move away from the UNFCCC may be particularly objectionable to those in the global South. Like other UN institutions, the UNFCCC works by consensus; each state has an equal say. The last two COPs, hosted by the petro-states Azerbaijan and the United Arab Emirates, have raised questions about whether countries must meet certain criteria to be eligible to host the COP. Yet introducing such a requirement violates the sacrosanct tenet of sovereign equality that underpins the United Nations. It is unclear which choice would do further damage to the UNFCCC's legitimacy (and to which nations).

At the same time, it is clear that maintaining the status quo of centering climate policy on the UNFCCC is also *hurting* developing countries—and running out the clock. Governments have been at this for thirty years, with little to show for it. And the stark reality of climate change is that those who are least responsible for causing it will bear the brunt of its effects. Thus, while a shift away from the UNFCCC may undercut some of the influence that developing nations have over climate policy, sticking with it will cause damage and death.

———

It's never an easy time to work on climate change, but this moment feels especially difficult. Trump's presidency is already proving disastrous for the climate and for US policy, and it will have reverberating effects on global politics, including climate cooperation. Many are looking to subnational and local action and grassroots organizing as a way to continue making progress in the face of the gross negligence of world leaders. Growing support for the Fossil Fuel Non-Proliferation Treaty also provides a glimmer of optimism.[44] Some governments are demanding that we address the elephant in the room: the supply of fossil fuels. These are laudable and important efforts—ones that keep us sane and hopeful in the face of fear and the onslaught of bad news.

My hope is that this book has offered other ways to maintain sanity and develop a new approach to climate politics. The proposals set forth here are not a panacea, but they are a step—a feasible step—in the right direction. We *can* approach climate politics in a different way, one that pivots away from technocratic debates about managing tons toward cultivating the political power of winners in the energy transition. Trade and finance institutions should be the central institutions to govern the climate. The Paris Agreement can provide transparency and track our collective progress, but we should not expect it to be the locus of global efforts to decarbonize.

Ultimately, global climate policy focused on assets can fix the failures of the Paris Agreement and build a more just and egalitarian society. Constraining fossil asset owners and investing in green asset owners can set us on a real path of decarbonization, one that cultivates lasting political support for the energy transition. We can end the fossil fuel era and create a more humane world.

NOTES

1. We Are Not All in This Together

1. United Nations Framework Convention on Climate Change, n.d.

2. Copernicus 2024a.

3. As of December 2024, 147 countries had made some form of net zero pledge. See Net Zero Tracker, "Data Tracker," https://zerotracker.net/.

4. Copernicus 2024b; UN Environment Programme 2023.

5. Colgan, Green, and Hale 2021.

6. Semieniuk et al. 2022.

7. Kelsey 2018; Kupzok and Nahm 2024.

8. Hale 2020.

9. Kupzok and Nahm 2024.

10. Oreskes and Conway 2011; Supran and Oreskes 2017.

11. Brulle 2018; Brulle and Downie 2022; Mildenberger 2020; Stokes 2020.

12. Green 2014.

13. Dauvergne and Lister 2015; Green et al. 2021; Si et al. 2023.

14. Abraham-Dukuma 2021.

15. Bergman 2018.

16. European Investment Bank 2020.

17. Bare, Colgan, and Gard-Murray 2025.

18. Bain 2025.

19. See, for example, Badgley et al. 2021; Gill-Wiehl, Kammen, and Haya 2024; Probst et al. 2024.

20. Intergovernmental Panel on Climate Change 2023.

21. Abdulla et al. 2020.

22. Clement et al. 2021; International Organization for Migration 2021.

23. Lyons 2022.

24. Frost 2023.

25. Chenet, Ryan-Collins, and van Lerven 2021.

26. Newman and Noy 2023.

27. O'Hara and Jones 2023.

28. Ibid.

29. Frank 2023.

30. International Renewable Energy Agency and International Labor Organization 2023.

31. Ibid.

32. Green, Hale, and Colgan 2019. On the challenges of a "just transition," see Lim, Aklin, and Frank 2023.

33. Some argue that we should consider those who *benefit* from climate change, such as farmers in relatively cold climes who may enjoy longer growing seasons. These arguments

ignore the larger context of climate change (overall losses outweigh benefits) and longer time frames (temperature increases above a certain level will be increasingly damaging). Given the many risks of climate change, only some of which may set off positive feedbacks, I reject beneficiaries of climate change as an analytically relevant category.

34. Green, Hale, and Colgan 2019.

35. Colgan, Green, and Hale 2021, 7.

36. This view of full decarbonization assumes that carbon capture and storage does not become a scalable, viable solution to climate change. Clearly, the jury is still out on this.

37. Semieniuk et al. 2022.

38. NewClimate Institute, Oxford Net Zero, Energy & Climate Intelligence Unit, and Data-Driven EnviroLab 2023.

39. Green et al. 2021.

40. Colgan, Green, and Hale 2021; Vormedal and Meckling 2024.

41. Of course, some governments are also huge fossil asset owners, but as I note later, I do not take on the question of petro-states in this book.

42. Oreskes and Conway 2011; Skjaerseth and Skodvin 2003.

43. Brulle and Downie 2022; Stokes 2020.

44. Oreskes and Conway 2011; Supran and Oreskes 2017, 2021.

45. Stokes 2020, 6.

46. GRAIN 2018; Lazarus, McDermid, and Jacquet 2021.

47. Sherrington, Carlile, and Healy 2023.

48. InfluenceMap 2018.

49. International Energy Agency 2021a.

50. International Energy Agency, n.d.

51. Hanto et al. 2022.

52. International Energy Agency, n.d.

53. World Bank 2024.

54. Green 2021c.

55. Baer 2016; Botrel et al. 2024.

56. Boffey 2021.

57. Kaminski 2024.

58. Paddison 2017.

59. Sabin Center for Climate Change Law 2022.

60. Keohane 2015.

61. Brulle and Downie 2022.

62. International Renewable Energy Agency 2023.

63. Davis et al. 2012; Kelley and Simmons 2015.

64. Barnett, Pevehouse, and Raustiala 2021.

65. Swyngedouw 2010, 219, 220.

66. Lohmann 2008, 364.

67. Victor 2011.

68. Unruh 2000.

69. Rosenbloom et al. 2020.

70. Green 2021a.

71. Saez and Zucman 2019, 111.

72. Galaz et al. 2018.

73. Jorgenson et al. 2016; Oxfam and Stockholm Environment Institute 2020; Padilla and Serrano 2006.

74. Bergin and Bousso 2020.

75. Tienhaara and Cotula 2020.

76. Tienhaara et al. 2023. See also https://investmentpolicy.unctad.org/investment-dispute
-settlement/cases/800/rockhopper-v-italy

77. International Center for Settlement of Investment Disputes 2024b.

78. Erickson and Lazarus 2014.

79. Tienhaara and Cotula 2020, table 1.

80. Tienhaara 2018.

81. US Department of Commerce 2024.

82. Bermel et al. 2024.

83. European Commission 2020.

84. Beckley 2023; Johnson 2022.

85. Nahm 2021; Tucker 2019b.

86. See, for example, Agarwal and Narain 1991; Sultana 2022; Táíwò 2022.

87. Colgan, Green, and Hale 2021.

88. Green 2021c. I should note that a more recent analysis comes up with much more generous estimates, but it suffers from some shortcomings. See Döbbeling-Hildebrandt et al. 2024.

89. Werksman 1998.

90. Green 2014, 2018.

91. United Nations Framework Convention on Climate Change 2015.

92. As reported by Net Zero Tracker (https://zerotracker.net/) as of November 25, 2024.

2. How Asset Revaluation Drives Existential Politics

1. See also Green 2021b.

2. Keohane 1984.

3. Stoddard et al. 2021.

4. World Meterological Organization 1979.

5. Barrett 2003; Bernauer 2013; Falkner, Stephan, and Vogler 2010; Keohane and Victor 2011,
2016; Nordhaus 2015; Ostrom 2010b; Sandler 2004; Stavins 2011; Stern 2007; Thompson 2006;
Victor 2011.

6. Barrett 1994, 878.

7. Barrett 2003, 2007.

8. Abbott et al. 2000; Green 2014, chap. 3; Green and Colgan 2013.

9. Aldy and Stavins 2007, 1.

10. Aldy and Stavins 2007; Nordhaus 1994; Schelling 1997. For the Stern Review, see Stern
2007.

11. Bernauer 2013, 2; Busby and Urpelainen 2020; Hovi, Sprinz, and Underdal 2009; Michaelowa and Michaelowa 2012; Thompson 2010; Tingley and Tomz 2014. Some important
exceptions include Gerlach and Rayner (1988) and Levin et al. (2012).

12. Keohane and Oppenheimer 2016, 143.

13. See, for example, Aklin and Mildenberger 2020; Bayer and Genovese 2020; Cheon and
Urpelainen 2013; Genovese 2019; Hughes and Urpelainen 2015; Kim et al. 2016; Lachapelle and
Paterson 2013; Meckling 2015; Mildenberger 2020.

14. Bernauer 2013, 134.

15. Thompson 2010.

16. Ostrom 2010a.

17. Bernstein and Hoffmann 2019.

18. Ibid., 923.

19. Intergovernmental Panel on Climate Change 1990, 61, 85.

20. Intergovernmental Panel on Climate Change 1995, 17.

21. Ibid., 47.

22. Global Climate Coalition 1992, 1.

23. Technically, it's a national carbon price, but provinces are allowed to have their own carbon pricing schemes, provided they comply with the minimum federal criteria.

24. Tasker 2021.

25. Corkal and Gass 2020.

26. Isachsen and Gylfason 2022.

27. UK Government 2019.

28. International Energy Agency 2021b.

29. Allegretti 2023.

30. Brulle 2023; Edwards et al. 2023; Ekberg et al. 2022.

31. Genovese 2020; Raymond 2020.

32. Green et al. 2021; Li, Trencher, and Asuka 2022.

33. Dauvergne and LeBaron 2014; Green 2024.

34. Avant, Finnemore, and Sell 2010; Cutler, Haufler, and Porter 1999; Green 2014.

35. Green 2014.

36. Ecosystem Marketplace 2024.

37. Badgley et al. 2021; Gill-Wiehl, Kammen, and Haya 2024; Kreibich and Hermwille 2021; Schneider 2009; Wara 2007.

38. For a prescient critique of the multilateral climate regime, see Victor 2001. On climate as a distributional problem, see note 13.

39. Ciplet, Roberts, and Khan 2015; Gardiner 2013; Healy and Barry 2017.

40. Agarwal and Narain 1991, 1.

41. Huber 2022; Klein 2014; Malm 2016; Mitchell 2013; Paterson 2021b; Wainwright and Mann 2018.

42. Malm 2016.

43. Huber 2022.

44. Klein 2014.

45. Newell and Paterson 2010, 9.

46. Hausfather 2024.

47. Paterson 2021.

48. European Environment Agency 2024.

49. I thank Danny Cullenward for this point and the terminology.

50. Becker-Olsen and Potucek 2013, 1318, as quoted in Nemes et al. 2022.

51. Green et al. 2021.

52. Kelsey 2018.

53. Colgan, Green, and Hale 2021.

54. Kelsey 2018.

55. Victor, Geels, and Sharpe 2019.

56. Kupzok and Nahm 2024.

57. Kupzok and Nahm 2024, 2.

58. Riofrancos 2020.

59. Colgan, Green, and Hale 2021.

60. Neville 2020.

61. Ørsted, n.d.

62. Brulle 2023.

63. Brulle, Roberts, and Spencer 2024.

64. Bousso 2023.

65. Bearak and Plumer 2023.

66. Stokes 2020.

67. Kelly, McNally, and Stephens 2024.

68. Javeline 2014.

69. Keohane 2015, 21.

70. Fiorino 2022; Sinha et al. 2023.

71. Intergovernmental Negotiating Committee for a Framework Convention on Climate Change 1991.

72. See, for example, Borenstein and Ritter 2015; Weikmans and Roberts 2019.

73. Cheon and Urpelainen 2013; Breetz, Mildenberger, and Stokes 2018.

74. Christophers 2022.

75. Mazzucato 2015.

76. Mildenberger 2020.

77. Tooze 2022.

78. Rodrik 2019.

79. Allan 2019.

80. Colgan and Keohane 2017.

81. Paterson 2021b.

82. Bergquist, Mildenberger, and Stokes 2020.

83. Mildenberger 2020.

3. The Limits of Carbon Pricing

1. See, for example, Stern 2007.

2. Rabe 2018, xvii.

3. World Bank 2024.

4. Green 2021c.

5. Stokes and Mildenberger 2020.

6. Syed 2021.

7. Paterson 2012.

8. Cullenward and Victor 2020, chap. 8.

9. Hepburn 2006.

10. World Bank 2022b.

11. Asian Development Bank 2018.

12. Dorsch, Flachsland, and Kornek 2020.

13. Green 2017.

14. Green 2008, 2014; Goldemberg 1998.

15. For a brief history of the climate regime, see Green 2021b.

16. Schwarze 2000.

17. Werksman 1998.

18. United Nations Framework Convention on Climate Change 2001.

19. Cullenward, Badgley, and Chay 2023; Goldemberg 1998; Werksman 1998.

20. Gillenwater 2012.

21. Cullenward, Badgley, and Chay 2023, 1085.

22. Grubb, Vrolijk, and Brack 1999, chap. 3.

23. Victor 2001, 26.

24. This section is based on findings from Green 2021, with previously unpublished updates.

25. Best, Burke, and Jotzo 2020.

26. Rafaty, Dolphin, and Pretis 2020.

27. Murray and Rivers 2015.

28. Pretis 2022, 117.

29. World Bank, n.d.

30. Meckling 2011.

31. Genovese and Tvinnereim 2019.

32. Bayer and Aklin 2020.

33. Döbbeling-Hildebrandt et al. 2024.

34. Döbbeling-Hildebrandt and his colleagues (2024) also use machine learning to identify relevant studies and extract the key data from each study, including the size of the policy's effect on reductions, statistical significance, and so on.

35. In theory, weighting each variable controls for these variations, but one critique of this approach is that it is difficult to account for this heterogeneity. See Hinne et al. 2020.

36. Chan and Morrow 2019; Fell and Maniloff 2018; Yan 2021; Zhou and Huang 2021.

37. Burtraw et al. 2023.

38. See, for example, Badgley et al. 2021.

39. Martin and Saikawa 2017; Wara 2014.

40. *Reuters* 2024.

41. Douenne and Fabre 2020; Mildenberger et al. 2022.

42. Genovese and Tvinnereim 2019; Kennard 2020; Meckling 2011; Raymond 2020. This is not to dismiss the inter- and intrasectoral variation in industry preferences on carbon pricing.

43. Mildenberger 2020; Crowley 2017.

44. Congress.gov 2009.

45. International Carbon Action Partnership 2022.

46. Yoder 2021.

47. Yoder 2019.

48. World Bank 2022a, 5.

49. Dayal 2024; Major 2023.

50. International Energy Agency 2020.

51. World Bank 2023a.

52. Genovese and Tvinnereim 2019.

53. This has been the case since 2008. In the pilot phase of the EU-ETS, each member state set its own cap.

54. Vogel and Kagan 2004.

55. European Commission 2003.

56. Since allowances could not be carried forward to the subsequent phase, their value fell to zero by the end of the first period.

57. Skjaerseth and Wettestad 2016, chap. 6; Wettestad and Jevnaker 2016, chap. 3.

58. Wettestad and Jevnaker 2016; Cullenward and Victor 2020, 42.

59. Gronwald and Hintermann 2015.

60. Bayer and Aklin 2020; Ranson and Stavins 2016.

61. European Commission, n.d.-.

62. International Energy Agency 2020.

63. Ellerman 2015; see also Dolphin 2022.

64. International Carbon Action Partnership 2022.

65. Gronwald and Hintermann 2015.

66. European Commission 2014.

67. European Parliament and Council 2015.

68. European Commission 2015.

69. Borghesi et al. 2023; Richstein, Chappin, and de Vries 2015.

70. Prices quoted from the Intercontinental Exchange.

71. AB 1279, passed in 2022, stipulates that California be net zero by 2045 and that 85 percent of the push toward that goal come from emissions reductions (as opposed to offsets).

72. Meckling and Nahm 2021.

73. Ibid. On insulation and climate policy, see Meckling et al. 2022.

74. Bang, Victor, and Andresen 2017.

75. California Air Resources Board 2020.

76. Burtraw et al. 2022.

77. Petek 2020.

78. Prete, Tyagi, and Xu 2023.

79. Pauer 2018.

80. Burtraw et al. 2023; Mastrandrea, Inman, and Cullenward 2020; Petek 2023a, 2023b.

81. Petek 2023a.

82. California Air Resources Board, n.d.-a.

83. Stiglitz and Stern 2017.

84. California Air Resources Board, n.d.-b.

85. Haya 2018.

86. Haya et al. 2020.

87. Ibid.

88. Badgley et al. 2022; Randazzo, Gordon, and Hamburg 2023.

89. Stapp et al. 2023.

90. Badgley et al. 2022.

91. Petek 2023a.

92. Huber and Shipan 2002.

93. Legislative Analyst's Office 2010.

94. For IEMAC annual reports, see California Environmental Protection Agency. n.d.

95. Wang, Carpenter-Gold, and So 2022, 31.

96. Green 2017.

97. World Bank 2024.

98. Ibid.

99. World Bank, n.d.

100. Federal Office for the Environment 2024.

101. Klik Foundation 2024.

102. La Hoz Theuer, Schneider, and Broekhoff 2019.

103. Mehling, Metcalf, and Stavins 2018b.

104. Keohane, Petsonk, and Hanafi 2017; Mehling, Metcalf, and Stavins 2018a; Ranson and Stavins 2016.

105. Green, Sterner, and Wagner 2014.

106. Green 2017.

107. Ibid.; Cullenward and Victor 2020.

108. Green, Sterner, and Wagner 2014.

109. Ibid.

110. Financial Accountability Office of Ontario 2018.

111. European Union 2023, paras. 21–22.

112. Mehling et al. 2019, 476.

113. Cosbey et al. 2012.

114. Ibid.; Mehling and Ritz 2020.

115. European Parliament and Council 2023a, para. 60.

116. European Parliament and Council 2023b.

117. Mehling et al. 2022.

118. Cullenward and Victor 2020.

119. Parry, Black, and Zhunussova 2022.

120. Ibid.

4. Carbon Offsets Are Fatally Flawed

1. United Nations Framework Convention on Climate Change 2024b.
2. Cullenward, Badgley, and Chay 2023.
3. Victor and House 2004; Victor, House, and Joy 2005.
4. Ecosystem Marketplace 2024, author's calculations based on table 3.
5. World Bank 2022b.
6. United Nations Framework Convention on Climate Change 2024b.
7. Green 2014, chap. 3.
8. Deacon and Murphy 1997.
9. Ecosystem Marketplace 2021.
10. Ecosystem Marketplace 2024.
11. Ibid.; Ecosystem Marketplace 2023, author's calculations based on table 7.
12. International Civil Aviation Organization 2019a, 240.
13. International Civil Aviation Organization 2020.
14. Kreibich and Hermwille 2021, 942.
15. United Nations Framework Convention on Climate Change 2024a.
16. United Nations Framework Convention on Climate Change 2022b, para. 75.
17. World Bank 2024.
18. World Bank 2023a.
19. Cames et al. 2016; Lund 2010; Schneider 2011; Wara 2006, 2007.
20. Badgley et al. 2021; Haya et al. 2023; Stapp et al. 2023.
21. Badgley et al. 2021; Haya et al. 2020; Lee et al. 2013.
22. Probst et al. 2024.
23. Clean Development Mechanism 2022.
24. Cecco 2023.
25. United Nations Framework Convention on Climate Change 2021.
26. Cullenward, Badgley, and Chay 2023; Romm 2023.
27. Ecosystem Marketplace 2023.
28. Schneider, Kollmuss, and Lazarus 2015.
29. Green 2013.
30. World Bank 2023a.
31. Rennert et al. 2022.
32. Ricke et al. 2018.
33. Tol 2019.
34. Greenfield 2024.
35. Blake 2023.
36. Twidale 2023.
37. Wenzel 2024.
38. For information, see the ICROA website at https://icroa.org/.
39. Cullenward and Victor 2020, 27.
40. This is not to dismiss the important issues of local communities that often don't receive their fair share of benefits or are adversely affected by some projects. Sadly, their objections are often overshadowed by larger, more powerful interests that push projects through. There *are* losers in the use of offsets, but they are not politically powerful enough to transform the status quo. Bumpus and Liverman 2008; Lyons and Westoby 2014.
41. Cullenward and Victor 2020, 94.
42. Budinis and Lo Re 2023.
43. Paterson and Stripple 2012.
44. Green 2010.

45. Green 2014, 6.

46. United Nations Framework Convention on Climate Change 2018; Ecosystem Marketplace 2024.

47. For example, see International Civil Aviation Organization 2019b for submissions to CORSIA's technical advisory board.

48. Ecosystem Marketplace 2023, author's calculations based on table 7.

49. Gold Standard 2023.

50. Salway 2021.

51. Bennett 2023.

52. Abbott, Green, and Keohane 2016; Green and Hadden 2021; Stroup and Wong 2018.

53. Ecosystem Marketplace 2024.

54. United Nations Framework Convention on Climate Change 2022b, annex 1, para. 75(a).

55. Schneider 2011.

56. Cullenward, Badgley, and Chay 2023; Cullenward and Victor 2020; Romm 2023.

57. Lakhani 2023.

58. Bloomberg NEF 2023.

59. Ibid., emphasis added.

60. Chay et al. 2022.

61. Blanchard, Anderegg, and Haya 2024.

62. Verra, n.d.

63. Romero 2023.

64. Greenfield and Kimeu 2023.

65. Greenfield 2023a, 2023b; Temple and Song 2021.

66. Amazon Watch 2021.

5. Net Zero: An Elaborate Distraction

1. Buck 2021.

2. As reported by Net Zero Tracker (https://zerotracker.net/) as of October 22, 2024.

3. Green 2024.

4. For an excellent political analysis of science-based targets, see Trexler and Schendler 2015.

5. Fankhauser et al. 2022, 15.

6. Lund et al. 2023.

7. Nakićenović and John 1991.

8. Foran and Crane 2002; Schneider et al. 2001; Waldheim and Carpentieri 2000.

9. IPCC 2000.

10. Potter et al. 2001; Prentice et al. 2000; Rödenbeck et al. 2003; Veenendaal et al. 2004; Waldheim and Carpentieri 2000.

11. Green and Reyes 2023; Rogelj et al. 2021.

12. Intergovernmental Panel on Climate Change 2007, 218.

13. Intergovernmental Panel on Climate Change 2023.

14. Fankhauser et al. 2022.

15. Science-Based Targets Initiative 2024b, 10.

16. Fankhauser et al. 2022, 16.

17. Intergovernmental Panel on Climate Change 2023, 11.

18. As reported by Net Zero Tracker (https://zerotracker.net/companies/jbs-com-0753) as of September 4, 2024.

19. Intergovernmental Panel on Climate Change 2023.

20. ExxonMobil 2022.

21. ExxonMobil 2024.

22. Intergovernmental Panel on Climate Change 1994.

23. Marland 2008; Marland, Hamal, and Jonas 2009.

24. Guan et al. 2012.

25. Intergovernmental Panel on Climate Change 2019.

26. Yona, Cashore, and Bradford 2022.

27. Davis and Caldeira 2010; Kander et al. 2015; Peters 2008; Peters and Hertwich 2008.

28. Peters 2008.

29. Ritchie, Rosado, and Roser 2023.

30. Green 2010, 2014.

31. Green 2014, chap. 6.

32. International Organization for Standardization 2018.

33. Walmart, n.d.

34. Gillenwater 2023.

35. Ibid.

36. World Business Council for Sustainable Development and World Resources Institute 2004.

37. Gillenwater 2023, 14.

38. Condon 2023, 1924.

39. Ibid., 1932–33.

40. See Greenhouse Gas Protocol (n.d.) for a list of some LCA datasets.

41. Gillenwater 2023, 6.

42. Ibid., 14.

43. CDP 2023; Hertwich and Wood 2018.

44. All figures in the paragraph are from Day et al. 2023.

45. Gillenwater 2023.

46. Bjørn et al. 2022; Brander, Gillenwater, and Ascui 2018; Hamburger 2019.

47. Gillenwater 2023.

48. Ibid.

49. This section is based on Green et al. 2024 and builds on the work of the Net Zero Tracker (www.zerotracker.net).

50. There are over seven hundred firms in the dataset, since some firms enter the Fortune 500 and some leave over the time period of the data.

51. United Nations Framework Convention on Climate Change 2022a.

52. Green, Hale, and Arceo 2024.

53. See, for example, Green 2014; Meckling 2011; Genovese and Tvinnereim 2019.

54. Coding details are available in Green et al. 2023. Each of the eight attributes is weighted equally in our index.

55. Fankhauser et al. 2022; New Climate Institute and Carbon Market Watch 2023; Rogelj et al. 2021.

56. Fankhauser et al. 2022.

57. Trencher et al. 2024.

58. See, for example, Allen et al. 2020; High-Level Expert Group on the Net Zero Emissions Commitments of Non-State Entities 2022; New Climate Institute and Carbon Market Watch 2023.

59. See also Fankhauser et al. 2022.

60. Fransen 2015.

61. McGivern et al. 2022.

62. Green 2014.

63. International Organization for Standardization 2022.

64. McGivern et al. 2022.

65. Race to Zero Expert Peer Review Group 2022.
66. Ibid., 15.
67. McGivern et al. 2022.
68. Aronczyk, McCurdy, and Russill 2024.
69. Day et al. 2022, 26.
70. Bjørn, Lloyd, and Matthews 2021; Rekker et al. 2022.
71. Hodgson 2022; Morse 2023.
72. Baue 2021.
73. Science Based Targets initiative 2024a.
74. Science Based Targets initiative 2023.
75. High-Level Expert Group on the Net Zero Emissions Commitments of Non-State Entities 2022, 16, 19, 34.
76. As reported by Net Zero Tracker (https://zerotracker.net/), as of April 27, 2024.
77. Bodansky 2016.
78. US Securities and Exchange Commission 2024; see also Condon 2023, 1924.
79. Dias and Axelsson 2023.
80. High-Level Expert Group on the Net Zero Emissions Commitments of Non-State Entities 2022, 22.
81. Ibid., 16.
82. Green and Auld 2017. Legal scholars often refer to this as "incorporation by reference"; see, for example, Bremer 2013.
83. Dias and Axelsson 2023.
84. US Securities and Exchange Commission 2024.
85. Ibid.
86. European Parliament and Council 2022.
87. Dias and Axelsson 2023.
88. Green 2014.
89. See, for example, its offset protocols for urban forestry and livestock projects at California Air Resources Board, n.d.-c.
90. Overdevest and Zeitlin 2014.
91. Bell and Hindmoor 2012.
92. Hale 2021.
93. Greenfield and Harvey 2024.
94. Science Based Targets Initiative 2024c.
95. Green and Denniss 2018.
96. Rogelj et al. 2019.
97. Green and Denniss 2018.
98. Newell and Simms 2020.
99. NewClimate Institute, Oxford Net Zero, Energy & Climate Intelligence Unit, and Data-Driven EnviroLab 2023. There is no updated data in its 2024 report.
100. Green et al. 2021, 2053 and figure 6.

6. Hit 'em Where It Hurts: Constraining Fossil Asset Owners

1. Brulle 2018, 2023; Brulle and Downie 2022; Franta 2021a, 2021b; Oreskes and Conway 2011; Supran and Oreskes 2017.
2. Khalfan et al. 2023, vii.
3. Ibid.
4. Ibid, xii.
5. Klein 2014.

6. Boute 2023; Cima 2021; Gallagher and Kozul-Wright 2022; Tienhaara 2009; Tienhaara and Cotula 2020; Verbeek 2023.

7. I use the term "regime" to refer to the sets of rules and institutions governing a given issue area. See Krasner 1983.

8. Unruh 2000.

9. Black 2023. This figure includes both implicit and explicit subsidies.

10. Ibid.

11. In 2010, a handful of countries created the group Friends of Fossil Fuel Subsidy Reform. The coalition currently has ten members: Costa Rica, Denmark, Ethiopia, Finland, the Netherlands, New Zealand, Norway, Sweden, Switzerland, and Uruguay. While more than thirty countries have "endorsed" the group's goals, they have not become members, further underscoring the extent to which subsidy reform has failed to garner political momentum.

12. Corporate Tax Haven Index 2024.

13. Zucman 2015, chap. 1.

14. Ibid.

15. Cobham 2017.

16. Tørsløv, Wier, and Zucman 2023; see also EU Tax Observatory 2024.

17. Cobham and Janský 2018.

18. Garcia-Bernardo and Janský 2024.

19. Jolly 2023.

20. Ibid.

21. Devereux et al. 2021.

22. Albertin et al. 2021.

23. Ibid., 6.

24. Ibid.

25. Kiezebrink 2023.

26. Kysar, forthcoming.

27. Friends of the Earth International, Bailout Watch, and Oxfam 2021.

28. Canadians for Tax Fairness 2017.

29. Cameron 2023.

30. BHP 2018.

31. Cobham and Janský 2018.

32. UN Environment Programme 2023.

33. Chancel 2022. For slightly different figures, see Oxfam 2023.

34. Oxfam and Institute for European Environmental Policy 2021.

35. Oxfam 2023; Wiedmann et al. 2020.

36. Murphy 2015.

37. Brulle 2023; Brulle and Downie 2022; Colgan, Green, and Hale 2021; Stokes 2020; Supran and Oreskes 2017.

38. Food and Agriculture Organization of the United Nations 2022.

39. Galaz et al. 2018.

40. Yayasan Auriga Nusantara et al. 2024.

41. In October 2024, the OECD opened a convention for signature that addresses the "subject to tax rule," which is the part of the Model Rules that addresses the corporate minimum tax.

42. Kysar, forthcoming.

43. Ibid.

44. Organization for Economic Cooperation and Development 2023.

45. Kamin and Kysar 2023.

46. Currently, the carve-out includes 10 percent of payroll and 8 percent of tangible assets.

47. Baraké et al. 2022.

48. Independent Commission for the Reform of International Corporate Taxation 2022, 2024.

49. PwC 2025.

50. Kysar 2023.

51. Organization for Economic Cooperation and Development 2023.

52. This is consistent with other types of global administrative law that utilize mechanisms such as transparency or participation to ensure administrative law–type mechanisms, since international rulemaking bodies do not have the same legal authority as domestic governments. See Kingsbury, Krisch, and Stewart 2005.

53. Herzfeld 2024.

54. Slaughter 2004.

55. UN General Assembly 2024a.

56. Ibid.

57. Drezner 2009; Kahraman, Kalyanpur, and Newman 2020.

58. Baraké et al. 2022.

59. Travers 2024.

60. Kysar 2023; EU Tax Observatory 2024.

61. Green and Denniss 2018.

62. The five key legal principles underpinning investment law are prohibition of expropriation without compensation, national treatment, most-favored nation treatment, fair and equitable treatment, and full protection and security.

63. Tienhaara and Cotula 2020.

64. Tienhaara 2009.

65. Bonnitcha, Poulsen, and Waibel 2017.

66. Cutler and Lark 2022.

67. Tienhaara 2018.

68. Tienhaara et al. 2023.

69. Arcuri, Tienhaara, and Pellegrini 2024.

70. International Center for Settlement of Investment Disputes 2019.

71. Tienhaara et al. 2022.

72. The claims were brought under the UN Commission on International Trade Law rather than ISDS. The same tenets of investment protections obtained.

73. Nottage 2024; Ranald 2024.

74. Di Salvatore 2021.

75. Bonnitcha and Brewin 2020.

76. Verbeek 2023.

77. International Center for Settlement of Investment Disputes 2024a, 2.

78. Neslen 2023.

79. Paine 2023; Ranjan 2023.

80. Tienhaara and Cotula 2020.

81. Tienhaara et al. 2022.

82. Ibid.

83. Tienhaara 2018.

84. Verbeek 2023.

85. Malo 2019.

86. International Energy Charter 2023.

87. Center for International Environmental Law 2022; Verbeek 2023.

88. Braoudakis, Craveia, and Baldon 2024.

89. Tienhaara et al. 2022.

90. Australia signed the ECT but never ratified it. It withdrew its signature in 2021.

91. Office of the United States Trade Representative 2020, chap. 14. However, Canada and Mexico can still invoke the ISDS under the Comprehensive and Progressive Agreement for Trans-Pacific Partnership (CPTPP).

92. Bulowski 2024.

93. Nottage 2024.

94. Organization for Economic Cooperation and Development, n.d.

95. Tienhaara et al. 2022.

96. Indeed, it is worth noting that just as some wealthy nations are withdrawing from the ECT, thirty-two other countries—primarily from Africa—have started the process to *join* the ECT. Ibid., appendix table S4.

97. Vogel and Kagan 2004.

98. See, for example, McNamara 2024.

99. HM Treasury 2024.

100. Mattli and Woods 2009.

7. Green Industrial Policy: Creating Green Asset Owners

1. Busby and Jensen 2023.

2. International Energy Agency 2021b.

3. International Renewable Energy Agency 2023, 25.

4. Mazzucato 2015.

5. Aiginger and Rodrik 2020, 189.

6. Allan, Lewis, and Oatley 2021, 3.

7. Lockwood 2015, 87, emphasis added.

8. Helveston and Nahm 2019.

9. Nahm 2021.

10. Bergquist, Mildenberger, and Stokes 2020.

11. Helveston and Nahm 2019.

12. Kelsey 2018; Kupzok and Nahm 2024; Victor, Geels, and Sharpe 2019.

13. Kupzok and Nahm 2024, 1204.

14. Kelsey 2018, 620.

15. Victor, Geels, and Sharpe 2019, 17.

16. Geels et al. 2017.

17. Organization for Economic Cooperation and Development 2024c, 8.

18. International Energy Agency 2021c.

19. Ibid.

20. International Energy Agency 2023a, 90.

21. International Energy Agency 2021c.

22. Canadian Manufacturing 2024.

23. Ontario Office of the Premier 2024.

24. There is disagreement about whether and how new mining activities will benefit the residents of the First Nations land where these extraction operations will be located.

25. Hercus 2024.

26. International Energy Agency 2022, 20.

27. International Energy Agency 2023a, chap. 4; see also International Energy Agency 2021c.

28. International Energy Agency 2023a, 74.

29. Lim, Aklin, and Frank 2023.

30. Hallegatte, Fay, and Vogt-Schilb 2013; Harrison, Martin, and Nataraj 2017.

31. Bistline et al. 2023.

32. Bermel et al. 2024.

33. Mildenberger and Stokes 2020.

34. Green 2021c; Mildenberger and Stokes 2020; Rabe 2018, chap. 2.

35. Meckling 2021.

36. Meckling 2021; Meckling et al. 2015.

37. Zhang et al. 2014.

38. Wang et al. 2012.

39. International Energy Agency 2022.

40. Hayley 2024.

41. Green 2017.

42. United Nations Framework Convention on Climate Change 2009.

43. Ibid.

44. Organization for Economic Cooperation and Development 2024a.

45. United Nations Framework Convention on Climate Change 2016, decision 1/CP21, para. 53.

46. United Nations Framework Convention on Climate Change 2024a, decision 1/CMA5, paras. 66–100.

47. Kumar and Ohri 2024.

48. Cash 2023.

49. Asian Development Bank 2024.

50. Beaufils et al. 2023.

51. There are other methods to produce green steel, such as molten oxide electrolysis or the use of green hydrogen as fuel. None are economically feasible at this point.

52. Falkner 2016; Nordhaus 2015; Stewart, Oppenheimer, and Rudyk 2013.

53. Unger and Quitzow 2024.

54. Sutton and Williams 2023.

55. Australian Government 2024.

56. Espa and Holzer 2023.

57. Ibid.

58. World Bank 2023b.

59. Organization for Economic Cooperation and Development 2024b.

60. Cosbey et al. 2017.

61. European Parliament and Council 2023c.

62. Industry for 2035, n.d.

63. Rodrik 2022.

64. Aronoff et al. 2019. Public services are not subject to the General Agreement on Trade in Services, the trade rules that govern services.

65. Kupzok and Nahm 2024, 1.

66. Meckling and Nahm 2018.

67. Bain 2024.

68. Meckling and Nahm 2021.

69. Bolet, Green, and González-Eguino 2024.

70. Breetz, Mildenberger, and Stokes 2018; Gaikwad, Genovese, and Tingley 2022.

71. The White House 2024.

72. Espiner 2024.

73. Blenkinsop 2024.

74. International Energy Agency 2023b.

75. Since the deadlock over the "Doha Round" of negotiations and the seemingly permanent hiatus of the WTO's enforcement arm, the Appellate Body, the authority of the WTO is waning.

76. Boudette 2024.
77. European Commission 2024.
78. Luo et al. 2024.
79. Nahm 2021.
80. Allan and Nahm 2024.
81. Ibid.
82. Bernstein and Hoffmann 2019.
83. Luo et al. 2024, 5.
84. Meckling et al. 2015.
85. Hale 2020.
86. Kennard 2020.
87. Gambhir et al. 2023.
88. Breetz, Mildenberger, and Stokes 2018; Stokes and Breetz 2018.
89. Breetz, Mildenberger, and Stokes 2018, 511–12.
90. Bernstein and Hoffmann 2019.
91. Alic 2020.
92. Juhász, Lane, and Rodrik 2023, 34.
93. International Renewable Energy Agency 2024, 50.
94. Fankhaeser, Sehllieir, and Stern 2008.
95. Tucker 2019a, 10.
96. Ibid., 35.
97. Sorbe, Gal, and Millot 2018.
98. Rodrik 2022.
99. Juhász, Lane, and Rodrik 2023.
100. Rodrik 2022, 15.
101. DeSombre and Barkin 2002; Esty 1994; Kamin and Kysar 2023; Lewis 2014, 2024.
102. Bacchus 2018.
103. New Zealand Foreign Affairs & Trade 2024.
104. Barrett 2007.
105. Meltzer 2013; Murray 2012.
106. European Parliament and Council 2024.
107. *Reuters* 2023.
108. Meyer and Tucker 2022, 117–18.
109. Klein 2014, 75–80.
110. Esty 1994.

8. The Future of Global Climate Policy

1. Colgan and Keohane 2017.
2. Enache 2024.
3. G20 Brazil 2024, 2024.
4. UN General Assembly 2024b, para. 10(b).
5. Thompson, Broude, and Haftel 2019.
6. Ibid., 859.
7. Mehranvar and Brauch 2024.
8. Peinhardt and Wellhausen 2016.
9. Bergquist, Mildenberger, and Stokes 2020; Green and Healy 2022.
10. Owen 2023.
11. Sorbe, Gal, and Millot 2018.
12. World Bank 2024.

13. Cullenward and Victor 2020.

14. Ibid., 97.

15. For similar arguments, see ibid., 140; see also Cullenward, Badgley, and Chay 2023; Green 2023.

16. United Nations Framework Convention on Climate Change 2024b, paras. 37–47.

17. Ibid.

18. There are no systematic cross-national data on the reliance on offsets.

19. Rumble 2024.

20. Green 2013; Kreibich and Hermwille 2021.

21. Green 2013.

22. European Commission, n.d.-b.

23. Green 2017.

24. See, for example, Ellis and Tirpak 2006; Ranson and Stavins 2016.

25. Canadian House of Commons 2024.

26. Woodside 2024.

27. Ibid.

28. Dias and Axelsson 2023.

29. Net Zero Regulation and Policy Hub 2023.

30. European Investment Bank 2020.

31. Broekhoff and Verkuijl 2024, 38.

32. United Nations Framework Convention on Climate Change 2024d.

33. Stiell 2024.

34. Bush 2024.

35. United Nations Framework Convention on Climate Change 2024c, para. 8.

36. Independent High-Level Expert Group on Climate Finance 2024. This figure is for emerging markets and developing countries, excluding China.

37. Club of Rome 2024.

38. UNFCCC 2024e.

39. Levy 1993.

40. These are the Global Environment Facility, the Green Climate Fund, the Special Climate Change Fund, the Least Developed Countries Fund, the Adaptation Fund, and the Fund for Responding to Loss and Damage.

41. Green Climate Fund 2024, 14.

42. Global Environmental Facility, n.d. The Special Climate Change Fund is administered by the Global Environmental Facility.

43. See article 7.1 of the Paris Agreement.

44. Newell and Simms 2020; see also Fossil Fuel Non-Proliferation Treaty Initiative, n.d.

REFERENCES

Abbott, Kenneth W., Jessica F. Green, and Robert O. Keohane. 2016. "Organizational Ecology and Institutional Change in Global Governance." *International Organization* 70 (2): 247–77.

Abbott, Kenneth, Robert Keohane, Andrew Moravcsik, Anne-Marie Slaughter, and Duncan Snidal. 2000. "The Concept of Legalization." *International Organization* 54 (3): 401–19.

Abdulla, Ahmed, Ryan Hanna, Kristen R. Schell, Oytun Babacan, and David G. Victor. 2020. "Explaining Successful and Failed Investments in US Carbon Capture and Storage Using Empirical and Expert Assessments." *Environmental Research Letters* 16 (1): 014036.

Abraham-Dukuma, Magnus C. 2021. "Dirty to Clean Energy: Exploring 'Oil and Gas Majors Transitioning.'" *The Extractive Industries and Society* 8 (3): 100936.

Agarwal, Anil, and Sunita Narain. 1991. "Global Warming in an Unequal World: A Case of Environmental Colonialism." *Earth Island Journal* (Spring): 81–91. https://doi.org/10.1093/oso/9780199498734.003.0005.

Aiginger, Karl, and Dani Rodrik. 2020. "Rebirth of Industrial Policy and an Agenda for the Twenty-First Century." *Journal of Industry, Competition, and Trade* 20 (2): 189–207.

Aklin, Michaël, and Matto Mildenberger. 2020. "Prisoners of the Wrong Dilemma: Why Distributive Conflict, Not Collective Action, Characterizes the Politics of Climate Change." *Global Environmental Politics* 20 (4): 4–27.

Albertin, Georgina, Boriana Yontcheva, Dan Devlin, Hilary Devine, Marc Gerard, Irena Suljagic, et al. 2021. "Tax Avoidance in Sub-Saharan Africa's Mining Sector." Washington, DC: World Bank (September 28). https://www.imf.org/en/Publications/Departmental-Papers-Policy-Papers/Issues/2021/09/27/Tax-Avoidance-in-Sub-Saharan-Africas-Mining-Sector-464850 (accessed November 15, 2023).

Aldy, Joseph E., and Robert N. Stavins, eds. 2007. *Architectures for Agreement: Addressing Global Climate Change in the Post- Kyoto World*. Cambridge: Cambridge University Press.

Alic, John A. 2020. "Endless Industrial Policy." *Issues in Science and Technology* 37 (1): 48–55.

Allan, Bentley, Joanna I. Lewis, and Thomas Oatley. 2021. "Green Industrial Policy and the Global Transformation of Climate Politics." *Global Environmental Politics* 21 (4): 1–19.

Allan, Bentley B., and Jonas Nahm. 2024. "Strategies of Green Industrial Policy: How States Position Firms in Global Supply Chains." *American Political Science Review* 119 (1): 420–34.

Allan, Jen Iris. 2019. "Dangerous Incrementalism of the Paris Agreement." *Global Environmental Politics* 19 (1): 4–11.

Allegretti, Aubrey. 2023. "Rishi Sunak Defies Critics and Presses On with Net Zero U-turn." *The Guardian*, September 21. https://www.theguardian.com/politics/2023/sep/21/rishi-sunak-defends-climate-policy-u-turn-despite-criticism (accessed October 30, 2024).

Allen, Myles, Kaya Axelsson, Ben Caldecott, Thomas Hale, Cameron Hepburn, Eli Mitchell-Larson, et al. 2020. "The Oxford Principles for Net Zero Aligned Carbon Offsetting 2020."

Oxford: University of Oxford, Smith School of Enterprise and the Environment (September). https://www.smithschool.ox.ac.uk/sites/default/files/2022-01/Oxford-Offsetting-Principles-2020.pdf.

Amazon Watch. 2021. "Statement: Offsets Don't Stop Climate Change." October 6. https://amazonwatch.org/news/2021/1006-statement-offsets-dont-stop-climate-change (accessed October 22, 2024).

Arcuri, Alessandra, Kyla Tienhaara, and Lorenzo Pellegrini. 2024. "Investment Law v. Supply-Side Climate Policies: Insights from *Rockhopper v. Italy* and *Lone Pine v. Canada*." *International Environmental Agreements: Politics, Law and Economics* 24 (1): 193–216.

Aronczyk, Melissa, Patrick McCurdy, and Chris Russill. 2024. "Greenwashing, Net-Zero, and the Oil Sands in Canada: The Case of Pathways Alliance." *Energy Research & Social Science* 112: 103502.

Aronoff, Kate, Alyssa Battistoni, Daniel Aldana Cohen, Thea Riofrancos, and Naomi Klein. 2019. *A Planet to Win: Why We Need a Green New Deal*. London: Verso.

Asian Development Bank (ADB). 2018. "The Korea Emissions Trading Scheme: Challenges and Emerging Opportunities." Manila: ADB (November). https://www.adb.org/publications/korea-emissions-trading-scheme (accessed March 1, 2024).

———. 2024. *Asian Economic Integration Report 2024: Decarbonizing Global Value Chains*. Manila: ADB (February). http://dx.doi.org/10.22617/SGP240085-2 (accessed November 13, 2024).

Australian Government. 2024. "Australia's Carbon Leakage Review." Canberra: Department of Climate Change, Energy, the Environment, and Water. Updated November 14, 2024. https://www.dcceew.gov.au/climate-change/emissions-reduction/review-carbon-leakage.

Avant, Deborah D., Martha Finnemore, and Susan K. Sell, eds. 2010. *Who Governs the Globe?* Cambridge: Cambridge University Press.

Bacchus, James. 2018. *The Content of a WTO Climate Waiver*. Waterloo, CA: Center for International Governance Innovation (December 4). https://www.cigionline.org/publications/content-wto-climate-waiver/ (accessed February 8, 2024).

Badgley, Grayson, Freya Chay, Oriana S. Chegwidden, Joseph J. Hamman, Jeremy Freeman, and Danny Cullenward. 2022. "California's Forest Carbon Offsets Buffer Pool Is Severely Undercapitalized." *Frontiers in Forests and Global Change* 5: 930426.

Badgley, Grayson, Jeremy Freeman, Joseph J. Hamman, Barbara Haya, Anna T. Trugman, William R. L. Anderegg, and Danny Cullenward. 2021. "Systematic Over-Crediting in California's Forest Carbon Offsets Program." *Global Change Biology* 28 (4): 1200–1203.

Baer, Hans A. 2016. "The Nexus of the Coal Industry and the State in Australia: Historical Dimensions and Contemporary Challenges." *Energy Policy* 99: 194–202.

Bain, Charles. 2025. "The Political Economy of Convertible Firms: Electric Utilities and Automakers in Climate Politics." PhD diss., University of Toronto.

Bang, Guri, David G. Victor, and Steinar Andresen. 2017. "California's Cap-and-Trade System: Diffusion and Lessons." *Global Environmental Politics* 17 (3): 12–30.

Baraké, Mona, Paul-Emmanuel Chouc, Theresa Neef, and Gabriel Zucman. 2022. "Revenue Effects of the Global Minimum Tax under Pillar Two." *Intertax* 50 (10): 689–710. https://doi.org/10.54648/taxi2022074.

Bare, Fiona, Jeff Colgan, and Alexander Gard-Murray 2025. "Has the Paris Climate Agreement Changed Corporate Behavior?" Unpublished manuscript.

Barnett, Michael N., Jon C. W. Pevehouse, and Kal Raustiala, eds. 2021. *Global Governance in a World of Change*. Cambridge: Cambridge University Press.

Barrett, Scott. 1994. "Self-Enforcing International Environmental Agreements." *Oxford Economic Papers* 46 (special issue, October): 878–94.

———. 2003. *Environment and Statecraft: The Strategy of Environmental Treaty-Making.* Oxford: Oxford University Press.

———. 2007. *Why Cooperate? The Incentive to Supply Global Public Goods.* Oxford: Oxford University Press.

Baue, Bill. 2021. "Formal Complaint/Request to Address Concerns over Science Based Targets Governance." Medium, February 15. https://bbaue.medium.com/formal-complaint-science -based-targets-conflicts-of-interest-f8199407ac10 (accessed February 17, 2025).

Bayer, Patrick, and Michaël Aklin. 2020. "The European Union Emissions Trading System Reduced CO_2 Emissions Despite Low Prices." *Proceedings of the National Academy of Sciences* 117 (16): 8804–12. https://www.pnas.org/content/early/2020/04/01/1918128117 (accessed April 9, 2020).

Bayer, Patrick, and Federica Genovese. 2020. "Beliefs about Consequences from Climate Action under Weak Climate Institutions: Sectors, Home Bias, and International Embeddedness." *Global Environmental Politics* 20 (4): 28–50.

Bearak, Max, and Brad Plumer. 2023. "In the End, an Oil Man Won a Climate Summit Deal on Moving Away from Oil." *New York Times,* December 13. https://www.nytimes.com/2023 /12/13/climate/sultan-al-jaber-cop28.html (accessed October 30, 2024).

Beaufils, Timothé, Hauke Ward, Michael Jakob, and Leonie Wenz. 2023. "Assessing Different European Carbon Border Adjustment Mechanism Implementations and Their Impact on Trade Partners." *Communications Earth & Environment* 4 (1): 1–9.

Becker-Olsen, Karen, and Sean Potucek. 2013. "Greenwashing." In *Encyclopedia of Corporate Social Responsibility,* edited by Samuel O. Idowu, Nicholas Capaldi, Liangrong Zu, and Ananda Das Gupta. Berlin: Springer.

Beckley, Michael. 2023. "Delusions of Détente: Why America and China Will Be Enduring Rivals." *Foreign Affairs,* August 22. https://www.foreignaffairs.com/united-states/china -delusions-detente-rivals (accessed February 23, 2024).

Bell, Stephen, and Andrew Hindmoor. 2012. "Governance without Government? The Case of the Forest Stewardship Council." *Public Administration* 90 (1): 144–59.

Bennett, Genevieve. 2023. "Ecosystem Marketplace and US Department of State to Assist Governments in Formulating Article 6 Carbon Markets Strategies." Ecosystem Marketplace. https:// www.ecosystemmarketplace.com/articles/ecosystem-marketplace-and-us-department-of -state-to-assist-governments-in-formulating-article-6-carbon-markets-strategies/ (accessed October 22, 2024).

Bergin, Tom, and Ron Bousso. 2020. "Special Report: How Oil Majors Shift Billions in Profits to Island Tax Havens." *Reuters,* December 9. https://www.reuters.com/article/global-oil-tax -havens-idUSKBN28J1IK (accessed December 27, 2020).

Bergman, Noam. 2018. "Impacts of the Fossil Fuel Divestment Movement: Effects on Finance, Policy, and Public Discourse." *Sustainability* 10 (7): 2529.

Bergquist, Parrish, Matto Mildenberger, and Leah C. Stokes. 2020. "Combining Climate, Economic, and Social Policy Builds Public Support for Climate Action in the US." *Environmental Research Letters* 15 (5): 054019.

Bermel, Lily, Ryan Cummings, Brian Deese, Michael Delgado, Leandra English, Yeric Garcia, et al. 2024. "Tallying the Two-Year Impact of the Inflation Reduction Act." New York and Cambridge, MA: Rhodium Group and MIT Center for Energy and Environmental Policy Research, August 7. https://www.cleaninvestmentmonitor.org/reports/tallying-the-two -year-impact-of-the-inflation-reduction-act (accessed November 11, 2024).

Bernauer, Thomas. 2013. "Climate Change Politics." *Annual Review of Political Science* 16 (1): 421–48.

Bernstein, Steven, and Matthew Hoffmann. 2019. "Climate Politics, Metaphors, and the Fractal Carbon Trap." *Nature Climate Change* 9 (12): 919–25.

Best, Rohan, Paul J. Burke, and Frank Jotzo. 2020. "Carbon Pricing Efficacy: Cross-Country Evidence." *Environmental and Resource Economics* 77: 69–94. https://doi.org/10.1007/s10640-020-00436-x (accessed August 13, 2020).

BHP. 2018. "BHP Settles Longstanding Transfer Pricing Dispute." Melbourne: BHP (November 19). https://www.bhp.com/news/media-centre/releases/2018/11/bhp-settles-longstanding-transfer-pricing-dispute (accessed October 28, 2024).

Bistline, John, Geoffrey Blanford, Maxwell Brown, Dallas Burtraw, Maya Domeshek, Jamil Farbes, et al. 2023. "Emissions and Energy Impacts of the Inflation Reduction Act." *Science* 380 (6652): 1324–27.

Bjørn, Anders, Shannon M. Lloyd, Matthew Brander, and H. Damon Matthews. 2022. "Renewable Energy Certificates Threaten the Integrity of Corporate Science-Based Targets." *Nature Climate Change* 12 (6): 539–46.

Bjørn, Anders, Shannon Lloyd, and Damon Matthews. 2021. "From the Paris Agreement to Corporate Climate Commitments: Evaluation of Seven Methods for Setting 'Science-Based' Emission Targets." *Environmental Research Letters* 16 (5): 054019.

Black, Simon, Antung A. Liu, Ian W. H. Parry, and Nate Vernon-Lin. 2023. "IMF Fossil Fuel Subsidies Data: 2023 Update." Washington, DC: IMF (August 24). https://www.imf.org/en/Publications/WP/Issues/2023/08/22/IMF-Fossil-Fuel-Subsidies-Data-2023-Update-537281 (accessed April 27, 2024).

Blake, Heidi. 2023. "The Great Cash-for-Carbon Hustle." *The New Yorker*, October 16. https://www.newyorker.com/magazine/2023/10/23/the-great-cash-for-carbon-hustle (accessed October 22, 2024).

Blanchard, Libby, William Anderegg, and Barbara K. Haya. 2024. "Instead of Carbon Offsets, We Need 'Contributions' to Forests." *Stanford Social Innovation Review* (January 31). https://ssir.org/articles/entry/forest-contributions-carbon-offsets (accessed September 9, 2024).

Blenkinsop, Philip. 2024. "EU Presses Ahead with Chinese EV Tariffs after Divided Vote." *Reuters*, October 4. https://www.reuters.com/business/autos-transportation/eu-governments-face-pivotal-vote-chinese-ev-tariffs-2024-10-04/ (accessed November 11, 2024).

Bloomberg NEF. 2023. "Long-Term Carbon Offsets Outlook 2023." Bloomberg Professional Services, July 18. https://www.bloomberg.com/professional/blog/long-term-carbon-offsets-outlook-2023/ (accessed October 2, 2023).

Bodansky, Daniel. 2016. "The Legal Character of the Paris Agreement." *Review of European, Comparative, & International Environmental Law* 25 (2): 142–50.

Boffey, Daniel. 2021. "Court Orders Royal Dutch Shell to Cut Carbon Emissions by 45% by 2030." *The Guardian*, May 26. https://www.theguardian.com/business/2021/may/26/court-orders-royal-dutch-shell-to-cut-carbon-emissions-by-45-by-2030 (accessed November 1, 2024).

Bolet, Diane, Fergus Green, and Mikel González-Eguino. 2024. "How to Get Coal Country to Vote for Climate Policy: The Effect of a 'Just Transition Agreement' on Spanish Election Results." *American Political Science Review* 118 (3): 1344–59.

Bonnitcha, Jonathan, and Sarah Brewin. 2020. "Compensation under Investment Treaties: What Are the Problems and What Can Be Done?" Manitoba: IISD (December 16). https://www.iisd.org/publications/compensation-under-investment-treaties (accessed January 22, 2023).

Bonnitcha, Jonathan, Lauge N. Skovgaard Poulsen, and Michael Waibel. 2017. *The Political Economy of the Investment Treaty Regime*. Oxford: Oxford University Press.

Borenstein, Seth, and Karl Ritter. 2015. "Billions in Climate Aid Pledges Have 'Wild West' Accounting." *AP News*, December 11. https://apnews.com/general-news-e0954e5e32c246b6852bf9693ec4cc6b (accessed April 23, 2024).

Borghesi, Simone, Michael Pahle, Grischa Perino, Simon Quemin, and Maximilian Willner. 2023. "The Market Stability Reserve in the EU Emissions Trading System: A Critical Review." *Annual Review of Resource Economics* 15 (1): 131–52.

Botrel, Clara Almeida, Saphira Rekker, Belinda Wade, and Syvannah Wilson. 2024. "Understanding the Lobbying Actions Taken by the Australian Renewable Energy Industry." *Journal of Cleaner Production* 434: 139674.

Boudette, Neal E. 2024. "Few Chinese Electric Cars Are Sold in US, but Industry Fears a Flood." *New York Times*, May 15. https://www.nytimes.com/2024/05/15/business/economy/china-electric-vehicles-biden-tariffs.html (accessed November 13, 2024).

Bousso, Ron. 2023. "Big Oil Doubles Profits in Blockbuster 2022." *Reuters*, February 8. https://www.reuters.com/business/energy/big-oil-doubles-profits-blockbuster-2022-2023-02-08/ (accessed October 30, 2024).

Boute, Anatole. 2023. "Investor Compensation for Oil and Gas Phase Out Decisions: Aligning Valuation Methods to Decarbonization." *Climate Policy* 23 (9): 1087–1100.

Brander, Matthew, Michael Gillenwater, and Francisco Ascui. 2018. "Creative Accounting: A Critical Perspective on the Market-Based Method for Reporting Purchased Electricity (Scope 2) Emissions." *Energy Policy* 112: 29–33.

Braoudakis, Nikos, Rosanne Craveia, and Clémentine Baldon. 2024. "Neutralising the ECT Sunset Clause Inter Se." *Foreign Investment Law Journal* 39 (2): 347–70. https://doi.org/10.1093/icsidreview/siae011.

Breetz, Hanna, Matto Mildenberger, and Leah Stokes. 2018. "The Political Logics of Clean Energy Transitions." *Business and Politics* 20 (4): 492–522.

Bremer, Emily S. 2013. "Incorporation by Reference in an Open-Government Age." *Harvard Journal of Law & Public Policy* 36 (1): 131–210.

Broekhoff, Derik, and Cleo Verkuijl. 2024. "Making Good on Financial Sector Net Zero Commitments: Building the Road to Policy." Stockholm: Stockholm Environment Institute (October). https://www.sei.org/publications/making-good-financial-net-zero-commitments-policy (accessed November 1, 2024).

Brulle, Robert J. 2018. "The Climate Lobby: A Sectoral Analysis of Lobbying Spending on Climate Change in the USA, 2000 to 2016." *Climatic Change* 149 (3): 289–303.

———. 2023. "Advocating Inaction: A Historical Analysis of the Global Climate Coalition." *Environmental Politics* 32 (2): 185–206.

Brulle, Robert, and Christian Downie. 2022. "Following the Money: Trade Associations, Political Activity, and Climate Change." *Climatic Change* 175 (3): 11.

Brulle, Robert, J. Timmons Roberts, and Miranda Spencer, eds. 2024. *Climate Obstruction across Europe.* Oxford: Oxford University Press.

Buck, Holly Jean. 2021. *Ending Fossil Fuels: Why Net Zero Is Not Enough.* Brooklyn: Verso.

Budinis, Sara, and Luca Lo Re. 2023. "Unlocking the Potential of Direct Air Capture: Is Scaling Up through Carbon Markets Possible?" Paris: IEA (May 11). https://www.iea.org/commentaries/unlocking-the-potential-of-direct-air-capture-is-scaling-up-through-carbon-markets-possible (accessed March 1, 2024).

Bulowski, Natasha. 2024. "TC Energy's $15-Billion Keystone XL Lawsuit Gets Thrown Out." *Canada's National Observer*, July 16. https://www.nationalobserver.com/2024/07/16/news/tc-energys-15-billion-keystone-xl-lawsuit-gets-thrown-out)accessed October 30, 2024).

Bumpus, Adam G., and Diana Liverman. 2008. "Accumulation by Decarbonization and the Governance of Carbon Offsets." *Economic Geography* 84 (2): 127–55.

Burtraw, Dallas, Danny Cullenward, Meredith Fowlie, Brian Holt, Katelyn Roedner Sutter, Ross Brown, and Shereen D'Souza. 2023. "2022 Annual Report of the Independent Emissions Market Advisory Committee." Independent Emissions Market Advisory Committee,

February 3. https://calepa.ca.gov/wp-content/uploads/sites/6/2023/02/2022-ANNUAL
-REPORT-OF-THE-INDEPENDENT-EMISSIONS-MARKET-ADVISORY
-COMMITTEE-2.pdf.

Burtraw, Dallas, Danny Cullenward, Meredith Fowlie, Katelyn Roedner Sutter, and Ross Brown. 2022. "2021 Annual Report of the Independent Emissions Market Advisory Committee." Independent Emissions Market Advisory Committee, February 4. https://calepa.ca.gov/wp -content/uploads/sites/6/2022/02/2021-IEMAC-Annual-Report.pdf.

Busby, Josh, and Nate Jensen. 2023. "Go Green Fast: Global Lessons for the Clean Energy Transition." Austin: University of Texas (February 24). https://sites.utexas.edu/gogreenfast/ (accessed November 13, 2024).

Busby, Joshua W., and Johannes Urpelainen. 2020. "Following the Leaders? How to Restore Progress in Global Climate Governance." *Global Environmental Politics* 20 (4): 99–121.

Bush, Rebecca. 2024. "'A Total Waste of Time': Why Papua New Guinea Pulled Out of COP29 and Why Climate Advocates Are Worried." *The Guardian*, November 7. https://www .theguardian.com/world/2024/nov/08/png-cop29-papua-new-guinea-un-climate -summit.

California Air Resources Board (CARB). 2020. "Review of Potential for Resource Shuffling in the Electricity Sector." Sacramento: CARB (February). https://ww2.arb.ca.gov/sites /default/files/cap-and-trade/guidance/resource_shuffling_faq.pdf.

———. n.d.-a. "Cap-and-Trade Program Data Dashboard." Sacramento: CARB. https://ww2 .arb.ca.gov/our-work/programs/cap-and-trade-program/program-data/cap-and-trade -program-data-dashboard (accessed October 26, 2024).

———. n.d.-b. "Compliance Offset Program." Sacramento: CARB. https://ww2.arb.ca.gov/our -work/programs/compliance-offset-program/about (accessed October 26, 2024).

———. n.d.-c. "Compliance Offset Protocols." Sacramento: CARB. https://ww2.arb.ca.gov /our-work/programs/compliance-offset-program/compliance-offset-protocols.

California Environmental Protection Agency. n.d. "Independent Emissions Market Advisory Committee." https://calepa.ca.gov/independent-emissions-market-advisory-committee/.

Cameron, Laura. 2023. "The Final Countdown: How Canada Can End Fossil Fuel Subsidies This Year." Winnipeg: International Institute for Sustainable Development (May 29). https://www.iisd.org/articles/insight/final-countdown-canada-fossil-fuel-subsidies (accessed October 28, 2024).

Cames, Martin, Ralph O. Harthan, Jürg Füssler, Michael Lazarus, Carrie M. Lee, Pete Erickson, and Randall Spalding-Fecher. 2016. *How Additional Is the Clean Development Mechanism? Analysis of the Application of Current Tools and Proposed Alternatives.* Frieburg: Institute for Applied Ecology (March 16). https://climate.ec.europa.eu/system/files/2017-04/clean _dev_mechanism_en.pdf.

Canadian House of Commons. 2024. "Fall Economic Statement Implementation Act, 2023" (Bill C-59). June 20. https://www.parl.ca/DocumentViewer/en/44-1/bill/C-59/royal-assent (accessed November 1, 2024).

Canadians for Tax Fairness. 2017. "Bay Street and Tax Havens: Curbing Corporate Canada's Addiction." Ottawa: Canadians for Tax Fairness (November). https://www.taxfairness.ca /sites/default/files/2022-07/report-bay-street-and-tax-havens-c4tf.pdf.

Canadian Manufacturing. 2024. "Honda Announces a Nearly $15B Investment in Ont., Plans to Build an EV Supply Chain." April 24. https://www.canadianmanufacturing.com /manufacturing/honda-announces-a-nearly-15b-investment-in-ont-plans-to-build-an-ev -supply-chain-299031/.

Cash, Joe. 2023. "China Urges EU to Ensure New Carbon Tax Complies with WTO Rules." *Reuters*, September 14. https://www.reuters.com/sustainability/china-urges-eu-ensure-new -carbon-tax-complies-with-wto-rules-2023-09-14/ (accessed November 13, 2024).

CDP. 2023. "CDP Technical Note: Relevance of Scope 3 Categories by Sector." Berlin: CDP (January 25). https://cdn.cdp.net/cdp-production/cms/guidance_docs/pdfs/000/003/504/original/CDP-technical-note-scope-3-relevance-by-sector.pdf (accessed May 1, 2024).

Cecco, Leyland. 2023. "Wildfires Turn Canada's Vast Forests from Carbon Sink into Super-Emitter." *The Guardian*, September 22. https://www.theguardian.com/world/2023/sep/22/canada-wildfires-forests-carbon-emissions (accessed October 27, 2024).

Center for International Environmental Law (CIEL). 2022. "A Backdoor for Fossil Fuel Protection: How Extending ECT Coverage to CCUS, Hydrogen, and Ammonia Will Lock in Oil and Gas." Washington, DC: CIEL (October). https://www.ciel.org/wp-content/uploads/2022/10/October-2022_CIEL_Briefing_A-Backdoor-for-Fossil-CIEL_brief_Fuel-Protection-How-Extending-ECT-Coverage-to-CCUS-Hydrogen-and-Ammonia-will-Lock-In-Oil-Gas-Oct-2022.pdf.

Chan, Nathan W., and John W. Morrow. 2019. "Unintended Consequences of Cap-and-Trade? Evidence from the Regional Greenhouse Gas Initiative." *Energy Economics* 80: 411422.

Chancel, Lucas. 2022. "Global Carbon Inequality over 1990–2019." *Nature Sustainability* 5 (11): 931–38.

Chay, Freya, Grayson Badgley, Kata Martin, Jeremy Freeman, Joseph J. Hamman, and Danny Cullenward. 2022. "Unpacking Ton-Year Accounting." (carbon)plan, January 31. https://carbonplan.org/research/ton-year-explainer (accessed November 4, 2024).

Chenet, Hugues, Josh Ryan-Collins, and Frank van Lerven. 2021. "Finance, Climate-Change, and Radical Uncertainty: Towards a Precautionary Approach to Financial Policy." *Ecological Economics* 183: 106957.

Cheon, Andrew, and Johannes Urpelainen. 2013. "How Do Competing Interest Groups Influence Environmental Policy? The Case of Renewable Electricity in Industrialized Democracies, 1989–2007." *Political Studies* 61 (4): 874–97.

Christophers, Brett. 2022. "Fossilised Capital: Price and Profit in the Energy Transition." *New Political Economy* 27 (1): 146–59.

Cima, Elena. 2021. "Retooling the Energy Charter Treaty for Climate Change Mitigation: Lessons from Investment Law and Arbitration." *Journal of World Energy Law & Business* 14 (2): 75–87.

Ciplet, David, J. Timmons Roberts, and Mizran Khan. 2015. *Power in a Warming World: The New Global Politics of Climate Change and the Remaking of Environmental Inequality.* Cambridge, MA: MIT Press.

Clean Development Mechanism. 2022. "ACM0009: Fuel Switching from Coal or Petroleum Fuel to Natural Gas, Version 5.0." Bonn: UNFCCC Secretariat. https://cdm.unfccc.int/methodologies/DB/CMUDOOMI7G7SYSDFXA75EIITKEVA4P.

Clement, Viviane, Kanta Kumari Rigaud, Alex de Sherbinin, Bryan Jones, Susana Adamo, Jacob Schewe, et al. 2021. "Groundswell Part 2: Acting on Internal Climate Migration." Washington, DC: World Bank (September 13). https://hdl.handle.net/10986/36248 (accessed October 30, 2024).

Club of Rome. 2024. "Open Letter on COP Reform to All States That Are Parties to the Convention," dated November 15, 2024, and addressed to Mr. Simon Stiell, Executive Secretary of the UNFCCC Secretariat, and UN Secretary-General António Guterres. https://www.clubofrome.org/cop-reform-2024/.

Cobham, Alex. 2017. "Tax Avoidance and Evasion—The Scale of the Problem." *Tax Justice Network Briefing* (November). https://www.taxjustice.net/wp-content/uploads/2017/11/Tax-dodging-the-scale-of-the-problem-TJN-Briefing.pdf (accessed February 20, 2025).

Cobham, Alex, and Petr Janský. 2018. "Global Distribution of Revenue Loss from Corporate Tax Avoidance: Re-estimation and Country Results." *Journal of International Development* 30 (2): 206–32.

Colgan, Jeff D., Jessica F. Green, and Thomas N. Hale. 2021. "Asset Revaluation and the Existential Politics of Climate Change." *International Organization* 75 (2): 586–610.

Colgan, Jeff D., and Robert O. Keohane. 2017. "The Liberal Order Is Rigged: Fix It Now or Watch It Wither." *Foreign Affairs* (May/June). https://www.foreignaffairs.com/articles/world/2017-04-17/liberal-order-rigged (accessed January 24, 2023).

Condon, Madison. 2023. "What's Scope 3 Good For?" *UC Davis Law Review* 56 (5): 1921–61.

Congress.gov. 2009. "H.R.2454: American Clean Energy and Security Act of 2009." 111th Cong. (2009–2010). https://www.congress.gov/bill/111th-congress/house-bill/2454/text?r=2&s=5#toc-H3E773A218135470AA982FE108029B88D (accessed April 25, 2024).

Copernicus. 2024a. "June 2024 Marks 12th Month of Global Temperature Reaching 1.5°C above Pre-Industrial." *Copernicus Climate Bulletins* (July 4). https://climate.copernicus.eu/copernicus-june-2024-marks-12th-month-global-temperature-reaching-15degc-above-pre-industrial (accessed November 26, 2024).

———. 2024b. "Surface Air Temperature for October 2024." https://climate.copernicus.eu/surface-air-temperature-october-2024.

Corkal, Vanessa, and Philip Gass. 2020. *Unpacking Canada's Fossil Fuel Subsidies.* Winnipeg: International Institute for Sustainable Development (December 11). https://www.iisd.org/articles/unpacking-canadas-fossil-fuel-subsidies-faq (accessed October 30, 2024).

Corporate Tax Haven Index. 2024. "The World's Biggest Enablers of Corporate Tax Abuse." https://cthi.taxjustice.net/en.

Cosbey, Aaron, Susanne Droege, Carolyn Fischer, Julia Reinaud, John Stephenson, Lutz Weischer, and Peter Wooders. 2012. "A Guide for the Concerned: Guidance on the Elaboration and Implementation of Border Carbon Adjustment." Winnipeg: International Institute for Sustainable Development (November 12). https://www.iisd.org/publications/report/guide-concerned-guidance-elaboration-and-implementation-border-carbon (accessed November 20, 2023).

Cosbey, Aaron, Peter Wooders, Richard Bridle, and Liesbeth Casier. 2017. "In with the Good, Out with the Bad: Phasing Out Polluting Sectors as Green Industrial Policy." In *Green Industrial Policy: Concept, Policies, Country Experiences,* edited by Tilman Altenburg and Claudia Assman. Geneva: UN Environment and German Development Institute.

Crowley, Kate. 2017. "Up and Down with Climate Politics 2013–2016: The Repeal of Carbon Pricing in Australia." *WIREs Climate Change* 8 (3): e458.

Cullenward, Danny, Grayson Badgley, and Freya Chay. 2023. "Carbon Offsets Are Incompatible with the Paris Agreement." *One Earth* 6 (9): 1085–88.

Cullenward, Danny, and David G. Victor. 2020. *Making Climate Policy Work.* Cambridge: Polity.

Cutler, A. Claire, Virginia Haufler, and Tony Porter, eds. 1999. *Private Authority and International Affairs.* Albany: State University of New York Press.

Cutler, A. Claire, and David Lark. 2022. "The Hidden Costs of Law in the Governance of Global Supply Chains: The Turn to Arbitration." *Review of International Political Economy* 29 (3): 719–48.

Dauvergne, Peter, and Genevieve LeBaron. 2014. *Protest Inc.: The Corporatization of Activism.* Cambridge: Polity.

Dauvergne, Peter, and Jane Lister. 2015. *Eco-Business: A Big-Brand Takeover of Sustainability.* Cambridge, MA: MIT Press.

Davis, Kevin, Angelina Fisher, Benedict Kingsbury, and Sally Engle Merry, eds. 2012. *Governance by Indicators: Global Power through Classification and Rankings.* Oxford: Oxford University Press.

Davis, Steven J., and Ken Caldeira. 2010. "Consumption-Based Accounting of CO_2 Emissions." *Proceedings of the National Academy of Sciences* 107 (12): 5687–92.

Day, Thomas, Silke Mooldijk, Frederic Hans, Sybrig Smit, Eduardo Posada, Reena Skribbe, et al. 2023. *Corporate Climate Responsibility Monitor 2023: Assessing the Transparency and Integrity of Companies' Emission Reduction and Net-Zero Targets.* Cologne and Brussels: NewClimate Institute and Carbon Market Watch (February). https://newclimate.org/sites/default/files /2023-04/NewClimate_CorporateClimateResponsibilityMonitor2023_Feb23.pdf.

Day, Thomas, Silke Mooldijk, Sybrig Smit, Eduardo Posada, Frederic Hans, Harry Fearne-hough, et al. 2022. *Corporate Climate Responsibility Monitor 2022: Assessing the Transparency and Integrity of Companies' Emission Reduction and Net-Zero Targets.* Cologne and Brussels: NewClimate Institute and Carbon Market Watch (February). https://newclimate.org/sites /default/files/2022-06/CorporateClimateResponsibilityMonitor2022.pdf.

Dayal, Pratyush. 2024. "Saskatchewan to Decide in February Whether to Remit Heating Gas Carbon Taxes to Ottawa, Duncan Says." *CBC News,* January 3. https://www.cbc.ca /news/canada/saskatoon/sask-will-make-a-decision-in-feb-on-carbon-tax-remittance-1 .7073375 (accessed October 18, 2024).

Deacon, Robert T., and Paul Murphy. 1997. "The Structure of an Environmental Transaction: The Debt-for-Nature Swap." *Land Economics* 73 (1): 1–24.

DeSombre, Elizabeth R., and J. Samuel Barkin. 2002. Turtles and Trade: The WTO's Acceptance of Environmental Trade Restrictions. *Global Environmental Politics* 2 (1): 12–18.

Devereux, Michael P., Alan J. Auerbach, Michael Keen, Paul Oosterhuis, Wolfgang Schön, and John Vella. 2021. *Taxing Profit in a Global Economy.* Oxford: Oxford University Press.

Dias, Lucilla, and Kaya Axelsson. 2023. "Net Zero Regulation Stocktake Report." Oxford: University of Oxford (November). https://netzeroclimate.org/wp-content/uploads/2023/11 /Net-Zero-Regulation-Stocktake-Report-November-2023.pdf.

Di Salvatore, Lea. 2021. "Investor-State Disputes in the Fossil Fuel Industry." Winnipeg: International Institute for Sustainable Development (December 31). https://www.iisd.org /publications/report/investor-state-disputes-fossil-fuel-industry (accessed October 23, 2024).

Döbbeling-Hildebrandt, Niklas, Klaas Miersch, Tarun Khanna, Marion Bachelet, Stephan B. Bruns, Max Callaghan, et al. 2024. "Systematic Review and Meta-Analysis of Ex-Post Evaluations on the Effectiveness of Carbon Pricing." *Nature Communications* (May 16). https:// doi.org/10.21203/rs.3.rs-2860638/v1 (accessed April 23, 2024).

Dolphin, Geoffroy. 2022. "World Carbon Pricing Database." Washington, DC: Resources for the Future (September 30). https://www.rff.org/publications/data-tools/world-carbon -pricing-database/ (accessed October 18, 2024).

Dorsch, Marcel J., Christian Flachsland, and Ulrike Kornek. 2020. "Building and Enhancing Climate Policy Ambition with Transfers: Allowance Allocation and Revenue Spending in the EU ETS." *Environmental Politics* 29 (5): 781–803.

Douenne, Thomas, and Adrien Fabre. 2020. "French Attitudes on Climate Change, Carbon Taxation, and Other Climate Policies." *Ecological Economics* 169: 106496.

Drezner, Daniel W. 2009. "The Power and Peril of International Regime Complexity." *Perspectives on Politics* 7 (1): 65–70.

Ecosystem Marketplace. 2021. "Market in Motion: The State of the Voluntary Carbon Markets 2021, Instalment 1." Washington, DC: Forest Trends Association. https://www .ecosystemmarketplace.com/publications/state-of-the-voluntary-carbon-markets-2021/.

———. 2023. "Paying for Quality: State of the Voluntary Carbon Markets 2023." Washington, DC: Forest Trends Association. https://www.ecosystemmarketplace.com/publications /state-of-the-voluntary-carbon-market-report-2023/.

———. 2024. "2024 State of the Voluntary Carbon Market (SOVCM): On the Path to Maturity." Washington, DC: Forest Trends Association. https://www.ecosystemmarketplace.com /publications/2024-state-of-the-voluntary-carbon-markets-sovcm/.

Edwards, Guy, Paul K. Gellert, Omar Faruque, Kathryn Hochstetler, Pamela D. McElwee, Prakash Kaswhan, et al. 2023. "Climate Obstruction in the Global South: Future Research Trajectories." *PLOS Climate* 2 (7): e0000241.

Ekberg, Kristoffer, Bernhard Forchtner, Martin Hultman, and Kirsti M. Jylhä. 2022. *Climate Obstruction: How Denial, Delay, and Inaction Are Heating the Planet.* London: Routledge.

Ellerman, A. Denny. 2015. "The EU ETS: What We Know and What We Don't Know." In *Emissions Trading as a Policy Instrument: Evaluation and Prospects*, edited by Marc Gronwald and Beat Hintermann. Cambridge, MA: MIT Press.

Ellis, Jane, and Dennis Tirpak. 2006. *"Linking GHG Emissions Trading Schemes and Markets."* Paris: OECD (October 25).

Enache, Cristina. 2024. "Windfall Profit Taxes in Europe, 2024." Brussels: Tax Foundation Europe (September 10). https://taxfoundation.org/data/all/eu/windfall-tax-europe-2024/.

Erickson, Peter, and Michael Lazarus. 2014. "Impact of the Keystone XL Pipeline on Global Oil Markets and Greenhouse Gas Emissions." *Nature Climate Change* 4 (9): 778–81.

Espa, Ilaria and Kateryna Holzer. 2023. "From Unilateral Border Carbon Adjustments to Cooperation in Climate Clubs: Rethinking Exclusion in Light of Trade and Climate Law Constraints." In *European Yearbook of International Economic Law 2022*, vol. 13, edited by J. Bäumler et al. Berlin: Springer.

Espiner, Tom. 2024. "China Hits Back at EU with Brandy Tax after Electric Car Tariffs." BBC, October 8. https://www.bbc.com/news/articles/cn8jz39xl19o (accessed November 13, 2024).

Esty, Daniel. 1994. *Greening the GATT: Trade, Environment, and the Future.* Washington, DC: Peterson Institute for International Economics.

EU Tax Observatory. 2024. *Global Tax Evasion Report 2024.* Paris: EU Tax Observatory. https://www.taxobservatory.eu//www-site/uploads/2023/10/global_tax_evasion_report_24.pdf (accessed February 20, 2024).

European Commission. 2003. "Directive 2003/87/EC of the European Parliament and of the Council of 13 October 2003 Establishing a Scheme for Greenhouse Gas Emission Allowance Trading within the Community and Amending Council Directive 96/61/EC." *Official Journal of the European Union* 46: 32–46.

——. 2014. "Commission Regulation (EU) No. 176/2014 of 25 February 2014 Amending Regulation (EU) No. 1031/2010 in Particular to Determine the Volumes of Greenhouse Gas Emission Allowances to Be Auctioned in 2013–20." *Official Journal of the European Union* 56: 11–13.

——. 2015. "Decision (EU) 2015/1814 of the European Parliament and of the Council of 6 October 2015 Concerning the Establishment and Operation of a Market Stability Reserve for the Union Greenhouse Gas Emission Trading Scheme and Amending Directive 2003/87/EC." *Official Journal of the European Union* 264: 1–5.

——. 2020. "The European Green Deal Investment Plan and Just Transition Mechanism Explained." Brussels: European Commission (January 13). https://ec.europa.eu/commission/presscorner/detail/en/qanda_20_24.

——. 2024. "Commission Proposal to Impose Tariffs on Imports of Battery Electric Vehicles from China Obtains Necessary Support from EU Member States." Brussels: European Union (October 4). https://ec.europa.eu/commission/presscorner/api/files/document/print/en/statement_24_5041/STATEMENT_24_5041_EN.pdf.

——. n.d.-a. "Use of International Credits." https://climate.ec.europa.eu/eu-action/eu-emissions-trading-system-eu-ets/use-international-credits_en.

——. n.d.-b. "Free Allocation." https://climate.ec.europa.eu/eu-action/eu-emissions-trading-system-eu-ets/free-allocation_en.

European Environment Agency. 2024. "Economic Losses from Weather- and Climate-Related Extremes in Europe." Copenhagen: European Environment Agency (October 14). https:// www.eea.europa.eu/en/analysis/indicators/economic-losses-from-climate-related (accessed October 30, 2024).

European Investment Bank (EIB). 2020. *EIB Group Climate Bank Roadmap 2021–2025*. Luxembourg: EIB (November). https://www.eib.org/attachments/thematic/eib_group_climate _bank_roadmap_en.pdf.

European Parliament and Council. 2015. "Decision (EU) 2015/1814 of the European Parliament and of the Council of 6 October 2015 concerning the establishment and operation of a market stability reserve for the Union greenhouse gas emission trading scheme and amending Directive 2003/87/EC." Brussels: Publications Office of the European Union (October 9). https://op.europa.eu/en/publication-detail/-/publication/478f05ef-a5d0-11ee -b164-01aa75ed71a1/language-en.

———. 2022. "Directive (EU) 2022/2464 of the European Parliament and of the Council of 14 December 2022 Amending Regulation (EU) No. 537/2014, Directive 2004/109/EC, Directive 2006/43/EC and Directive 2013/34/EU, as Regards Corporate Sustainability Reporting." Brussels: European Union (December 14). http://data.europa.eu/eli/dir/2022/2464/oj.

———. 2023a. "Regulation (EU) 2023/956 of the European Parliament and of the Council of 10 May 2023 Establishing a Carbon Border Adjustment Mechanism." Brussels: European Union (May 10). http://data.europa.eu/eli/reg/2023/956/oj.

———. 2023b. "Directive (EU) 2023/959 of the European Parliament and of the Council of 10 May 2023 Amending Directive 2003/87/EC Establishing a System for Greenhouse Gas Emission Allowance Trading within the Union and Decision (EU) 2015/1814 Concerning the Establishment and Operation of a Market Stability Reserve for the Union Greenhouse Gas Emission Trading System." Brussels: European Union (May 16). https://eur-lex.europa .eu/eli/dir/2023/959/oj/eng.

———. 2023c. "Regulation (EU) 2023/851 of the European Parliament and of the Council of 19 April 2023 Amending Regulation (EU) 2019/631 as Regards Strengthening the CO2 Emission Performance Standards for New Passenger Cars and New Light Commercial Vehicles in Line with the Union's Increased Climate Ambition." Brussels: European Union (April 19). http://data.europa.eu/eli/reg/2023/851/oj 2023a.

———. 2024. "Regulation (EU) 2024/1735 of the European Parliament and of the Council of 13 June 2024 on Establishing a Framework of Measures for Strengthening Europe's Net-Zero Technology Manufacturing Ecosystem and Amending Regulation (EU) 2018/1724 (Text with EEA Relevance)." Brussels: European Union (June 13). http://data.europa.eu/eli/reg /2024/1735/oj/eng (accessed October 3, 2024).

ExxonMobil. 2022. "ExxonMobil Announces Ambition for Net Zero Greenhouse Gas Emissions by 2050." January 18. https://corporate.exxonmobil.com/news/news-releases/2022 /0118_exxonmobil-announces-ambition-for-net-zero-greenhouse-gas-emissions-by-2050 (accessed October 22, 2024).

———. 2024. "2024 Advancing Climate Solutions." January 8. https://corporate.exxonmobil .com/-/media/global/files/advancing-climate-solutions/2024/2024-advancing-climate -solutions-report.pdf.

Falkner, Robert. 2016. "A Minilateral Solution for Global Climate Change? On Bargaining Efficiency, Club Benefits, and International Legitimacy." *Perspectives on Politics* 14 (1): 87–101.

Falkner, Robert, Hannes Stephan, and John Vogler. 2010. "International Climate Policy after Copenhagen: Towards a 'Building Blocks' Approach." *Global Policy* 1 (3): 252–62.

Fankhaeser, Samuel, Friedel Sehllieir, and Nicholas Stern. 2008. "Climate Change, Innovation, and Jobs." *Climate Policy* 8 (4): 421–29. https://www.tandfonline.com/doi/abs/10.3763 /cpol.2008.0513 (accessed October 8, 2024).

Fankhauser, Sam, Stephen M. Smith, Myles Allen, Kaya Axelsson, Thomas Hale, Cameron Hepburn, et al. 2022." The Meaning of Net Zero and How to Get It Right." *Nature Climate Change* 12 (1): 15–21.

Food and Agriculture Organization of the United Nations (FAO). 2022. *The State of the World's Forests 2022: Forest Pathways for Green Recovery and Building Inclusive, Resilient, and Sustainable Economies*. Rome: FAO. https://www.fao.org/3/cb9360en/cb9360en.pdf.

Federal Office for the Environment (FOEN). 2024. "Bilateral Climate Agreements." Bern: FOEN (updated May 14). https://www.bafu.admin.ch/bafu/en/home/topics/climate/info-specialists/climate--international-affairs/staatsvertraege-umsetzung-klimauebereinkommen-von-paris-artikel6.html.

Fell, Harrison, and Peter Maniloff. 2018. "Leakage in Regional Environmental Policy: The Case of the Regional Greenhouse Gas Initiative." *Journal of Environmental Economics and Management* 87: 1–23.

Financial Accountability Office of Ontario (FAO). 2018. "Cap and Trade: A Financial Review of the Decision to Cancel the Cap and Trade Program." Toronto: FAO (October 16). https://fao-on.org/wp-content/uploads/2024/08/Cap-and-Trade.pdf.

Fiorino, Daniel J. 2022. "Democracies, Authoritarians, and Climate Change: Do Regime Types Matter?" In *How Democracy Survives: Global Challenges in the Anthropocene*, edited by Michael Holm and R. S. Deese. Routledge.

Foran, Barney, and David Crane. 2002. "Testing the Feasibility of Biomass Based Transport Fuels and Electricity Generation in Australia." *Australian Journal of Environmental Management* 9 (2): 103–14.

Fossil Fuel Non-Proliferation Treaty Initiative. n.d. https://fossilfueltreaty.org/.

Frank, Thomas. 2023. "Climate Change Is Destabilizing Insurance Industry." *Scientific American*, March 23. https://www.scientificamerican.com/article/climate-change-is-destabilizing-insurance-industry/ (accessed February 28, 2024).

Fransen, Luc. 2015. "The Politics of Meta-Governance in Transnational Private Sustainability Governance." *Policy Sciences* 48 (3): 293–317.

Franta, Benjamin. 2021a. "Early Oil Industry Disinformation on Global Warming." *Environmental Politics* 30 (4): 663–68.

———. 2021b. "Weaponizing Economics: Big Oil, Economic Consultants, and Climate Policy Delay." *Environmental Politics* 31 (4): 555–75.

Friends of the Earth International, Bailout Watch, and Oxfam. 2021. "12 Guilty Fogeys: Big Oil's $86 Billion Offshore Tax Bonanza." Merrifield, VA: Friends of the Earth International. https://foe.org/wp-content/uploads/2021/09/FFS_12_Guilty_Fogeys_rd3.pdf.

Frost, Natasha. 2023. "No, 11,200 Climate Refugees Aren't Heading to Australia." *New York Times*, November 11. https://www.nytimes.com/2023/11/11/world/australia/tuvalu-climate.html (accessed October 30, 2024).

G20 Brazil 2024. 2024. "The Rio de Janeiro G20 Ministerial Declaration on International Tax Cooperation." https://www.gov.br/fazenda/pt-br/assuntos/g20/declaracoes/1-g20-ministerial-declaration-international-taxation-cooperation.pdf.

Gaikwad, Nikhar, Federica Genovese, and Dustin Tingley. 2022. "Creating Climate Coalitions: Mass Preferences for Compensating Vulnerability in the World's Two Largest Democracies." *American Political Science Review* 116 (4): 1165–83.

Galaz, Victor, Beatrice Crona, Alice Dauriach, Jean-Baptiste Jouffray, Henrik Österblom, and Jan Fichtner. 2018. "Tax Havens and Global Environmental Degradation." *Nature Ecology & Evolution* 2 (9): 1352–57.

Gallagher, Kevin P., and Richard Kozul-Wright. 2022. *The Case for a New Bretton Woods*. Cambridge: Polity.

Gambhir, Ajay, Shivika Mittal, Robin D. Lamboll, Neil Grant, Dan Bernie, Laila Gohar, et al. 2023. "Adjusting 1.5 Degree C Climate Change Mitigation Pathways in Light of Adverse New Information." *Nature Communications* 14 (1): 5117.

Garcia-Bernardo, Javier, and Petr Janský. 2024. "Profit Shifting of Multinational Corporations Worldwide." *World Development* 177: 106527.

Gardiner, Stephen M. 2013. *A Perfect Moral Storm: The Ethical Tragedy of Climate Change.* Oxford: Oxford University Press.

Geels, Frank W., Benjamin K. Sovacool, Tim Schwanen, and Steve Sorrell. 2017. "The Socio-Technical Dynamics of Low-Carbon Transitions." *Joule* 1 (3): 463–79.

Genovese, Federica. 2019. "Sectors, Pollution, and Trade: How Industrial Interests Shape Domestic Positions on Global Climate Agreements." *International Studies Quarterly* 63 (4): 819–36.

———. 2020. "Market Responses to Global Governance: International Climate Cooperation and Europe's Carbon Trading." *Business and Politics* 23 (1): 91–123.

Genovese, Federica, and Endre Tvinnereim. 2019. "Who Opposes Climate Regulation? Business Preferences for the European Emission Trading Scheme." *Review of International Organizations* 14 (3): 511–42.

Gerlach, Luther, and Steve Rayner. 1988. "Culture and the Common Management of Global Risks." *Practicing Anthropology* 10 (3/4): 15–18.

Gillenwater, Michael. 2012. "What Is Additionality? Part 1: A Long Standing Problem." Washington, DC: GHG Management Institute (January). https://ghginstitute.org/wp-content/uploads/2015/04/AdditionalityPaper_Part-1ver3FINAL.pdf.

———. 2023. "What Is Greenhouse Gas Accounting? Turning Away from LCA." Washington, DC: GHG Management Institute (September). https://ghginstitute.org/wp-content/uploads/2023/12/What-is-GHG-Accounting-Turning-Away-from-LCA-Installment-N-1-23.12.19.pdf.

Gill-Wiehl, Annelise, Daniel M. Kammen, and Barbara K. Haya. 2024. "Pervasive Over-Crediting from Cookstove Offset Methodologies." *Nature Sustainability* 7: 192–202.

Global Climate Coalition. 1992. "The Global Climate Coalition Recognizes US Leadership for Presenting National Plan in Geneva." Washington, DC: Global Climate Coalition (December 8). https://www.documentcloud.org/documents/5628884-GCC-1992-12-8-Recognizes-US-Leadership-for.html.

Global Environmental Facility (GEF). n.d. "Special Climate Change Fund—SCCF." Washington, DC: GEF. https://www.thegef.org/what-we-do/topics/special-climate-change-fund-sccf.

Goldemberg, Jose. 1998. "Overview." In *Issues and Options: The Clean Development Mechanism,* edited by Jose Goldemberg. New York: United Nations Development Programme.

Gold Standard. 2023. "Technical Support for Early Article 6 Activities." Châtelaine, Switzerland: Gold Standard (July 10). https://www.goldstandard.org/news/technical-support-early-article-6-activities (accessed October 27, 2024).

GRAIN. 2018. "Emissions Impossible: How Big Meat and Dairy Are Heating Up the Planet." Minneapolis: Institute for Agriculture & Trade Policy (July 18). https://www.iatp.org/emissions-impossible (accessed March 1, 2024).

Green Climate Fund. 2024. *Annual Report 2023.* Incheon, Korea: Green Climate Fund (September 23). https://www.greenclimate.fund/document/annual-report-2023 (accessed November 14, 2024).

Green, Fergus, and Richard Denniss. 2018. "Cutting with Both Arms of the Scissors: The Economic and Political Case for Restrictive Supply-Side Climate Policies." *Climatic Change* 150 (1): 73–87.

Green, Fergus, and Noel Healy. 2022. "How Inequality Fuels Climate Change: The Climate Case for a Green New Deal." *One Earth* 5 (6): 635–49.

Green, Jessica. 2008. "Delegation and Accountability in the Clean Development Mechanism: The New Authority of Non-State Actors." *Journal of International Law and International Relations* 4 (2): 21–55.

———. 2010. "Private Standards in the Climate Regime: The Greenhouse Gas Protocol." *Business and Politics* 12 (3): 1–37.

———. 2013. "Order Out of Chaos: Public and Private Rules for Managing Carbon." *Global Environmental Politics* 13 (2): 1–25.

———. 2014. *Rethinking Private Authority: Agents and Entrepreneurs in Global Environmental Governance.* Princeton, NJ: Princeton University Press.

———. 2017. "Don't Link Carbon Markets." *Nature News* 543 (7646): 484.

———. 2018. "From Green to REDD: Protean Power and the Politics of Carbon Sinks." In *Protean Power: Exploring the Uncertain and Unexpected in World Politics.* Cambridge: Cambridge University Press.

———. 2021a. "Beyond Carbon Pricing: Tax Reform Is Climate Policy." *Global Policy* 12 (3): 372–79.

———. 2021b. "Climate Change Governance: Past, Present and (Hopefully) Future." In *Global Governance in a World of Change*, edited by Michael N. Barnett, Jon C. W. Pevehouse, and Kal Raustiala. Cambridge: Cambridge University Press.

———. 2021c. "Does Carbon Pricing Reduce Emissions? A Review of Ex-Post Analyses." *Environmental Research Letters* 16 (4). http://iopscience.iop.org/article/10.1088/1748-9326/abdae9 (accessed February 16, 2021).

———. 2023. "The False Promise of Carbon Offsets." *Foreign Affairs*, November 20. https://www.foreignaffairs.com/world/false-promise-carbon-offsets (accessed May 28, 2024).

———. 2024. "The Climate Establishment and the Paris Partnerships." *Climatic Change* 177 (84).

Green, Jessica F., and Graeme Auld. 2017. "Unbundling the Regime Complex: The Effects of Private Authority." *Transnational Environmental Law* 6 (2): 259–84.

Green, Jessica F., and Jeff Colgan. 2013. "Protecting Sovereignty, Protecting the Planet: State Delegation to International Organizations and Private Actors in Environmental Politics." *Governance* 26 (3): 473–97.

Green, Jessica F., and Jennifer Hadden. 2021. "How Did Environmental Governance Become Complex? Understanding Mutualism between Environmental NGOs and International Organizations." *International Studies Review* 23 (4): 1792–1812.

Green, Jessica, Jennifer Hadden, Thomas Hale, and Paasha Mahdavi. 2021. "Transition, Hedge, or Resist? Understanding Political and Economic Behavior toward Decarbonization in the Oil and Gas Industry." *Review of International Political Economy* 29 (6): 2036–20.

Green, Jessica, Thomas Hale, and Aldrick Arceo. 2024. "The Net Zero Wave: Identifying Patterns in the Uptake and Robustness of National and Corporate Net Zero Targets 2015–2023." *Climate Policy* (online, September 28). https://doi.org/10.1080/14693062.2024.2405221.

Green, Jessica F., Thomas N. Hale, and Jeff D. Colgan. 2019. The Existential Politics of Climate Change." *Global Policy Journal* (February 21). https://www.globalpolicyjournal.com/blog/21/02/2019/existential-politics-climate-change (accessed May 27, 2022).

Green, Jessica F., and Raúl Salas Reyes. 2023. "The History of Net Zero: Can We Move from Concepts to Practice?" *Climate Policy* 23 (7): 901–15.

Green, Jessica F., Thomas Sterner, and Gernot Wagner. 2014. "A Balance of Bottom-Up and Top-Down in Linking Climate Policies." *Nature Climate Change* 4 (12): 1064–67.

Greenfield, Patrick. 2023a. "Revealed: More than 90% of Rainforest Carbon Offsets by Biggest Certifier Are Worthless, Analysis Shows." *The Guardian*, January 18. https://www

.theguardian.com/environment/2023/jan/18/revealed-forest-carbon-offsets-biggest -provider-worthless-verra-aoe (accessed December 5, 2023).

———. 2023b. "Carbon Credit Speculators Could Lose Billions as Offsets Deemed Worthless." *The Guardian*, August 24. https://www.theguardian.com/environment/2023/aug/24 /carbon-credit-speculators-could-lose-billions-as-offsets-deemed-worthless-aoe (accessed December 5, 2023).

Greenfield, Patrick. 2024. "Ex-Carbon Offsetting Boss Charged in New York with Multimillion-Dollar Fraud." *The Guardian*, October 4. https://www.theguardian.com/environment/2024 /oct/04/ex-carbon-offsetting-boss-kenneth-newcombe-charged-in-new-york-with -multimillion-dollar (accessed October 21, 2024).

Greenfield, Patrick, and Fiona Harvey. 2024. "Climate Target Organisation Faces Staff Revolt over Carbon-Offsetting Plan." *The Guardian*, April 11. https://www.theguardian.com /environment/2024/apr/11/climate-target-organisation-faces-staff-revolt-over-carbon -offsetting-plan-sbti (accessed November 4, 2024).

Greenfield, Patrick, and Caroline Kimeu. 2023. "Shell Signals Retreat from Carbon Offsetting." *The Guardian*, September 8. https://www.theguardian.com/environment/2023/sep/08 /shell-signals-retreat-from-carbon-offsetting (accessed October 22, 2024).

Greenhouse Gas Protocol. n.d. "Life Cycle Databases." https://ghgprotocol.org/life-cycle -databases.

Gronwald, Marc, and Beat Hintermann. 2015. "The EU ETS." In *Emissions Trading as a Policy Instrument*, edited by Marc Gronwald and Beat Hintermann. Cambridge, MA: MIT Press.

Grubb, Michael, with Christian Vrolijk and Duncan Brack. 1999. *The Kyoto Protocol: A Guide and Assessment*. London: Royal Institute of International Affairs.

Guan, Dabo, Zhu Liu, Yong Geng, Sören Lindner, and Klaus Hubacek. 2012. "The Gigatonne Gap in China's Carbon Dioxide Inventories." *Nature Climate Change* 2 (9): 672–75.

Hale, Thomas. 2020. "Catalytic Cooperation." *Global Environmental Politics* 20 (4): 73–98.

———. 2021. "Governing Net Zero: The Conveyor Belt" (policy memo). Oxford University, Blavatnik School of Government, November 29. https://www.bsg.ox.ac.uk/research /publications/governing-net-zero-conveyor-belt (accessed February 20, 2025).

Hallegatte, Stéphane, Marianne Fay, and Adrien Vogt-Schilb. 2013. "Green Industrial Policies: When and How." Policy Research Working Paper 6677. Washington, DC: World Bank (October). http://elibrary.worldbank.org/doi/book/10.1596/1813-9450-6677 (accessed February 22, 2024).

Hamburger, Ákos. 2019. "Is Guarantee of Origin Really an Effective Energy Policy Tool in Europe? A Critical Approach." *Society and Economy* 41 (4): 487–507.

Hanto, Jonathan, Akira Schroth, Lukas Krawielicki, Pao-yu Oei, and Jesse Burton. 2022. "The Political Economy of Energy and Climate Policy in South Africa." In *The Political Economy of Coal: Obstacles to Clean Energy Transitions*, edited by Michael Jakob and Jan Steckel. Abingdon: Routledge.

Harrison, Ann, Leslie A. Martin, and Shanthi Nataraj. 2017. "Green Industrial Policy in Emerging Markets." *Annual Review of Resource Economics* 9 (1): 253–74.

Hausfather, Zeke. 2024. "State of the Climate: 2023 Smashes Records for Surface Temperature and Ocean Heat." Carbon Brief, January 12. https://www.carbonbrief.org/state-of-the -climate-2023-smashes-records-for-surface-temperature-and-ocean-heat/ (accessed October 30, 2024).

Haya, Barbara K. 2018. "The Size of California's Carbon Offset Program." Berkeley: California Institute for Energy and Environment, June 12. https://gspp.berkeley.edu/assets/uploads /page/FACTSHEET-the-size-of-CAs-offset-program-Haya.pdf.

Haya, Barbara, Danny Cullenward, Aaron L. Strong, Emily Grubert, Robert Heilmayr, Deborah A. Sivas, and Michael Wara. 2020. "Managing Uncertainty in Carbon Offsets: Insights from California's Standardized Approach." *Climate Policy* 20 (9): 1112–26.

Haya, Barbara K., Samuel Evans, Letty Brown, Jacob Bukoski, Van Butsic, Bodie Cabiyo, et al. 2023. "Comprehensive Review of Carbon Quantification by Improved Forest Management Offset Protocols." *Frontiers in Forests and Global Change* 6: 958879.

Hayley, Andrew. 2024. "Explainer: China's Dominance in Wind Turbine Manufacturing." *Reuters*, April 10. https://www.reuters.com/business/energy/chinas-dominance-wind -turbine-manufacturing-2024-04-10/ (accessed May 1, 2024).

Healy, Noel, and John Barry. 2017. "Politicizing Energy Justice and Energy System Transitions: Fossil Fuel Divestment and a 'Just Transition.'" *Energy Policy* 108: 451–59.

Helveston, John, and Jonas Nahm. 2019. "China's Key Role in Scaling Low-Carbon Energy Technologies." *Science* 366 (6467): 794–96.

Hepburn, Cameron. 2006. "Regulation by Prices, Quantities, or Both: A Review of Instrument Choice." *Oxford Review of Economic Policy* 22 (2): 226–47.

Hercus, Catherine. 2024. "Passing [through] the Ring of Fire: Recent Developments." *Canadian Mining Journal*, March 5. https://www.canadianminingjournal.com/featured-article/passing -though-the-ring-of-fire-recent-developments/ (accessed November 11, 2024).

Hertwich, Edgar G., and Richard Wood. 2018. "The Growing Importance of Scope 3 Greenhouse Gas Emissions from Industry." *Environmental Research Letters* 13 (10): 104013.

Herzfeld, Mindy. 2024. "OECD Rulemaking, the APA, and *Chevron* Deference." Tax Notes, January 15. https://www.taxnotes.com/featured-analysis/oecd-rulemaking-apa-and -chevron-deference/2024/01/12/7j27m (accessed September 22, 2024).

High-Level Expert Group on the Net Zero Emissions Commitments of Non-State Entities (HLEG). 2022. "Integrity Matters: Net Zero Commitments by Businesses, Financial Institutions, Cities, and Regions: Report from the United Nations' High-Level Expert Group on the Net Zero Emissions Commitments of Non-State Entities." United Nations. https://www .un.org/sites/un2.un.org/files/high-level_expert_group_n7b.pdf.

Hinne, Max, Quentin F. Gronau, Don van den Bergh, and Eric-Jan Wagenmakers. 2020. "A Conceptual Introduction to Bayesian Model Averaging." *Advances in Methods and Practices in Psychological Science* 3 (2): 200–215.

HM Treasury. 2024. "Changes to the Energy (Oil and Gas) Profits Levy." GOV.UK, July 29. https://www.gov.uk/government/publications/july-statement-2024-changes-to-the -energy-oil-and-gas-profits-levy/changes-to-the-energy-oil-and-gas-profits-levy (accessed October 28, 2024).

Hodgson, Camilla. 2022. "Climate Scientists Criticise Corporate Emissions Oversight Body SBTi." *Financial Times*, November 1. https://www.ft.com/content/8efe3f48-1f00-4731-919b -e47f4d5f7c82 (accessed February 20, 2024).

Hovi, Jon, Detlef F. Sprinz, and Arild Underdal. 2009. "Implementing Long-Term Climate Policy: Time Inconsistency, Domestic Politics, International Anarchy." *Global Environmental Politics* 9 (3): 20–39.

Huber, John D., and Charles R. Shipan. 2002. *Deliberate Discretion? The Institutional Foundations of Bureaucratic Autonomy.* Cambridge: Cambridge University Press.

Huber, Matthew T. 2022. *Climate Change as Class War: Building Socialism on a Warming Planet.* London: Verso.

Hughes, Llewelyn, and Johannes Urpelainen. 2015. "Interests, Institutions, and Climate Policy: Explaining the Choice of Policy Instruments for the Energy Sector." *Environmental Science & Policy* 54: 52–63.

Independent Commission for the Reform of International Corporate Taxation (ICRICT). 2022. "It Is Time for a Global Asset Registry to Tackle Hidden Wealth." ICRICT,

April. https://www.icrict.com/wp-content/uploads/2023/10/ICRICTGARreportEN
.pdf.

———. 2024. "ICRICT Evaluation of the OECD/G2O Two-Pillar Solution." ICRICT, September 28. https://www.icrict.com/international-tax-reform/icrict-evaluation-of-the-oecd
-g20-two-pillar-solution-2/ (accessed October 28, 2024).

Independent High-Level Expert Group on Climate Finance (IHLEG). 2024. "Raising Ambition and Accelerating Delivery of Climate Finance." London: Grantham Research Institute
on Climate Change and the Environment, London School of Economics and Political
Science (November). https://www.lse.ac.uk/granthaminstitute/wp-content/uploads
/2024/11/Raising-ambition-and-accelerating-delivery-of-climate-finance_Executive
-summary.pdf.

Industry for 2035. n.d. https://industryfor2035.org/about.

InfluenceMap. 2018. "How the US Auto Industry Is Dismantling the World's Most Successful
Climate Policy." London: InfluenceMap (April). https://influencemap.org/report/How-the
-US-auto-industry-is-dismantling-the-US-s-most-successful-climate-change-policy
-5c079bd28ca4e219519afa0ae462db08 (accessed November 1, 2024).

Intergovernmental Negotiating Committee for a Framework Convention on Climate Change.
1991. "[V.] Insurance Mechanism" (addendum to the Consolidated Working Document).
A/AC.237/MISC.17/Add.9. Geneva: Intergovernmental Negotiating Committee for a
Framework Convention on Climate Change (December 19). https://unfccc.int/resource
/docs/1991/a/eng/misc17a09.pdf.

Intergovernmental Panel on Climate Change (IPCC). 1990. *Climate Change: The IPCC Scientific
Assessment*, edited by J. T. Houghton, G. J. Jenkins, and J. J. Ephraums. Cambridge: Cambridge University Press. https://archive.ipcc.ch/publications_and_data/publications_ipcc
_first_assessment_1990_wg1.shtml.

———. 1994. "IPCC Guidelines for National Greenhouse Gas Inventories." Geneva: IPCC
(updated in 1996). https://www.ipcc.ch/report/ipcc-guidelines-for-national-greenhouse
-gas-inventories/.

———. 1995. "Second Assessment: Climate Change 1995." Geneva: IPCC. https://archive.ipcc
.ch/pdf/climate-changes-1995/ipcc-2nd-assessment/2nd-assessment-en.pdf.

———. 2007. *Climate Change 2007: Mitigation of Climate Change: Contribution of Working Group
III to the Fourth Assessment Report of the Intergovernmental Panel on Climate Change*. Cambridge: Cambridge University Press. https://www.ipcc.ch/report/ar4/wg3/.

———. 2019. *2019 Refinement to the 2006 IPCC Guidelines for National Greenhouse Gas Inventories*. Geneva: IPCC. https://www.ipcc.ch/report/2019-refinement-to-the-2006-ipcc
-guidelines-for-national-greenhouse-gas-inventories/.

———. 2023. "Climate Change 2023: Synthesis Report: Contribution of Working Groups I, II,
and III to the Sixth Assessment Report of the IPCCC," 35–115. Geneva: IPCC. https://doi:
10.59327/IPCC/AR6-9789291691647.

International Carbon Action Partnership (ICAP). 2022. "EU Emissions Trading System (EU
ETS)." Berlin: ICAP. https://icapcarbonaction.com/system/files/ets_pdfs/icap-etsmap
-factsheet-43.pdf.

International Center for Settlement of Investment Disputes (ICSID). 2019. "In the Arbitration
Proceeding between Rockhopper Italia S.P.A., Rockhopper Mediterranean Ltd, and Rockhopper Exploration Plc: Decision on the Intra-EU Jurisdictional Objection." ICSID case
no. ARB/17/14, June 26. https://www.italaw.com/sites/default/files/case-documents
/italaw10646_0.pdf.

———. 2024a. *Klesch Group Holdings Limited and Raffinerie Heide GmbH v. Federal Republic of
Germany*. Case no. ARB/23/49. Washington, DC: ICSID. https://www.italaw.com/cases
/10931.

International Center for Settlement of Investment Disputes (ICSID). 2024b. *TC Energy Corporation and TransCanada Pipelines Limited v. United States of America (II)*. Case no. ARB/21/63. Washington, DC: ICSID. https://www.italaw.com/cases/9339 (accessed November 1, 2024).

International Civil Aviation Organization (ICAO). 2019a. *Destination Green: The Next Chapter: 2019 Environmental Report*. Montreal: ICAO. https://www.icao.int/environmental-protection/Documents/ICAO-ENV-Report2019-F1-WEB%20(1).pdf.

———. 2019b. "2019 TAB Assessment." Montreal: ICAO. https://www.icao.int/environmental-protection/CORSIA/Pages/TAB2019.aspx.

———. 2020. "Recommendations on CORSIA Eligible Units." Montreal: ICAO, Technical Advisory Board (January). https://www.icao.int/environmental-protection/CORSIA/Documents/TAB/TAB%202020/Excerpt_TAB_Report_Jan_2020_final.pdf.

International Energy Agency (IEA). 2020. "Implementing Effective Emissions Trading Systems: Lessons from International Experiences." Paris: IEA. https://iea.blob.core.windows.net/assets/2551e81a-a401-43a4-bebd-a52e5a8fc853/Implementing_Effective_Emissions_Trading_Systems.pdf.

———. 2021a. "Global EV Outlook: Policies to Promote Electric Vehicle Deployment." Paris: IEA. https://www.iea.org/reports/global-ev-outlook-2021/policies-to-promote-electric-vehicle-deployment (accessed March 1, 2024).

———. 2021b. "Net Zero by 2050." Paris: IEA (May). https://www.iea.org/reports/net-zero-by-2050.

———. 2021c. "Executive Summary." In *The Role of Critical Minerals in Clean Energy Transitions*. Paris: IEA (May). https://www.iea.org/reports/the-role-of-critical-minerals-in-clean-energy-transitions/executive-summary (accessed May 3, 2024).

———. 2022. *Special Report on Solar PV Global Supply Chains*. Paris: OECD Publishing (August 26). https://www.oecd-ilibrary.org/energy/special-report-on-solar-pv-global-supply-chains_9e8b0121-en (accessed May 1, 2024).

———. 2023a. "Clean Energy Supply Chains Vulnerabilities." In *Energy Technology Perspectives 2023*. Paris: IEA (January). https://www.iea.org/reports/energy-technology-perspectives-2023/clean-energy-supply-chains-vulnerabilities (accessed May 5, 2024).

———. 2023b. "Tracking Clean Energy Progress 2023." Paris: IEA (July). https://www.iea.org/reports/tracking-clean-energy-progress-2023#overview.

———. n.d. "South Africa." Paris: IEA. https://www.iea.org/countries/south-africa/coal.

International Energy Charter. 2023. A summary of the efforts to modernize the Energy Charter Treaty. https://www.energychartertreaty.org/modernisation-of-the-treaty/.

International Organization for Migration (IOM). 2021. *World Migration Report 2022*. Geneva: IOM. https://publications.iom.int/books/world-migration-report-2022 (accessed March 1, 2024).

International Organization for Standardization (IOS). 2018. "Part 1: Specification with Guidance at the Organization Level for Quantification and Reporting of Greenhouse Gas Emissions and Removals." Geneva: IOS. https://www.iso.org/standard/66453.html (accessed April 29, 2024).

———. 2022. "Net Zero Guidelines." In "ISO Net Zero Guidelines," IWA 42(2022). Geneva: IOS. https://www.iso.org/climate-change/embracing-net-zero#toc5.

International Renewable Energy Agency (IRENA). 2023. *World Energy Transitions Outlook 2023: 1.5°C Pathway*. Abu Dhabi: IRENA (June). https://www.irena.org/Publications/2023/Jun/World-Energy-Transitions-Outlook-2023 (accessed March 7, 2024).

———. 2024. *Geopolitics of the Energy Transition: Energy Security*. Abu Dhabi: IRENA. https://www.irena.org/Publications/2024/Apr/Geopolitics-of-the-energy-transition-Energy-security (accessed May 3, 2024).

International Renewable Energy Agency (IRENA) and International Labor Organization (ILO). 2023. *Renewable Energy and Jobs: Annual Review 2023*, 10th ed. Abu Dhabi: IRENA and ILO. https://www.irena.org/Publications/2023/Sep/Renewable-energy-and-jobs-Annual-review-2023 (accessed February 24, 2024).

Isachsen, Arne Jon, and Thorvaldur Gylfason. 2022. "Putting Oil Profits to Global Benefit." Washington, DC: International Monetary Fund (December). https://www.imf.org/en/Publications/fandd/issues/2022/12/POV-putting-oil-profits-to-global-benefit-isachsen-gylfason (accessed October 30, 2024).

Javeline, Debra. 2014. "The Most Important Topic Political Scientists Are Not Studying: Adapting to Climate Change." *Perspectives on Politics* 12 (2): 420–34.

Johnson, Ian. 2022. "Has China Lost Europe? How Beijing's Economic Missteps and Support for Russia Soured European Leaders." *Foreign Affairs*, June 10. https://www.foreignaffairs.com/articles/china/2022-06-10/has-china-lost-europe (accessed February 23, 2024).

Jolly, Jasper. 2023. "UK Lost Out on £2bn in Tax in 2021 as Big Tech Shifted Profits Abroad, Claim Campaigners." *The Guardian*, October 16. https://www.theguardian.com/business/2023/oct/16/uk-lost-out-on-2bn-in-tax-in-2021-as-big-tech-shifted-profits-abroad-claim-campaigners (accessed February 20, 2024).

Jorgenson, Andrew K., Juliet B. Schor, Kyle W. Knight, and Xiaorui Huang. 2016. "Domestic Inequality and Carbon Emissions in Comparative Perspective." *Sociological Forum* 31 (S1): 770–86.

Juhász, Réka, Nathan J. Lane, and Dani Rodrik. 2023. "The New Economics of Industrial Policy." Working paper 31538. Cambridge, MA: National Bureau of Economic Research (August). https://www.nber.org/papers/w31538 (accessed March 7, 2024).

Kahraman, Filiz, Nikhil Kalyanpur, and Abraham L. Newman. 2020. "Domestic Courts, Transnational Law, and International Order." *European Journal of International Relations* 26 (1, suppl.): 184–208.

Kamin, David, and Rebecca Kysar. 2023. "The Perils of the New Industrial Policy." *Foreign Affairs* (May/June). https://www.foreignaffairs.com/united-states/industrial-policy-china-perils (accessed October 3, 2023).

Kaminski, Isabella. 2024. "Shell Defeats Landmark Climate Ruling Ordering Cut in Carbon Emissions." *The Guardian*, November 12. https://www.theguardian.com/environment/2024/nov/12/shell-wins-appeal-against-court-ruling-ordering-cut-in-carbon-emissions (accessed November 20, 2024).

Kander, Astrid, Magnus Jiborn, Daniel D. Moran, and Thomas O. Wiedmann. 2015. "National Greenhouse-Gas Accounting for Effective Climate Policy on International Trade." *Nature Climate Change* 5 (5): 431–35.

Kelley, Judith G., and Beth A. Simmons. 2015. "Politics by Number: Indicators as Social Pressure in International Relations." *American Journal of Political Science* 59 (1): 55–70.

Kelly, Orla, Brenda McNally, and Jennie C. Stephens. 2024. "Climate Obstruction in Ireland: The Contested Transformation of an Agricultural Economy." In *Climate Obstruction across Europe*, edited by Robert Brulle, J. Timmons Roberts, and Miranda Spencer. Oxford: Oxford University Press.

Kelsey, Nina. 2018. "Industry Type and Environmental Policy: Industry Characteristics Shape the Potential for Policymaking Success in Energy and the Environment." *Business and Politics* 20 (4): 615–42.

Kennard, Amanda. 2020. "The Enemy of My Enemy: When Firms Support Climate Change Regulation." *International Organization* 74 (2): 187–221.

Keohane, N., A. Petsonk, and A. Hanafi. 2017. "Toward a Club of Carbon Markets." *Climatic Change* 144 (1): 81–95.

Keohane, Robert O. 1984. *After Hegemony: Cooperation and Discord in the World Political Economy*. Princeton, NJ: Princeton University Press.

———. 2015. "The Global Politics of Climate Change: Challenge for Political Science." *PS: Political Science & Politics* 48 (1): 19–26.

Keohane, Robert O., and Michael Oppenheimer. 2016. "Paris: Beyond the Climate Dead End through Pledge and Review?" *Politics and Governance* 4 (3): 142–51.

Keohane, Robert O., and David G. Victor. 2011. "The Regime Complex for Climate Change." *Perspectives on Politics* 9 (1): 7–23.

———. 2016. "Cooperation and Discord in Global Climate Policy." *Nature Climate Change* 6 (6): 570–75.

Khalfan, Ashfaq, Astrid Nilsson Lewis, Carlos Aguilar, Jacqueline Persson, Max Lawson, Nafkote Dabi, et al. 2023. *Climate Equality: A Planet for the 99%*. Oxford: Oxfam.

Kiezebrink, Vincent. 2023. "Tax Avoidance in Mozambique's Extractive Industries." Amsterdam: SOMO (July 21). https://www.somo.nl/the-treaty-trap/ (accessed October 28, 2024).

Kim, Sung Eun, Johannes Urpelainen, and Joonseok Yang. 2016. "Electric Utilities and American Climate Policy: Lobbying by Expected Winners and Losers." *Journal of Public Policy* 36 (2): 251–75.

Kingsbury, Benedict, Nico Krisch, and Richard B. Stewart. 2005. "The Emergence of Global Administrative Law." *Law and Contemporary Problems* 68 (15): 15–61.

Klein, Naomi. 2014. *This Changes Everything: Capitalism vs. the Climate*. Toronto: Knopf Canada.

Klik Foundation. 2024. "First Ever ITMOs for NDC Use." Zurich: Klik Foundation (January 8). https://www.klik.ch/en/news/news-article/first-ever-itmos-for-ndc-use (accessed November 7, 2024).

Krasner, Stephen D., ed. 1983. *International Regimes*. Ithaca, NY: Cornell University Press.

Kreibich, Nicolas, and Lukas Hermwille. 2021. "Caught in Between: Credibility and Feasibility of the Voluntary Carbon Market Post-2020." *Climate Policy* 21 (7): 939–57.

Kumar, Manoj, and Nikunj Ohri. 2024. "India Sees EU Carbon Tax Proposal as Unfair and Not Acceptable, Official Says." *Reuters*, July 29. https://www.reuters.com/world/india/india-sees-eu-carbon-tax-proposal-unfair-not-acceptable-official-says-2024-07-29/ (accessed November 13, 2024).

Kupzok, Nils, and Jonas Nahm. 2024. "The Decarbonization Bargain: How the Decarbonizable Sector Shapes Climate Politics." *Perspectives on Politics* (August 20): 1203–23.

Kysar, Rebecca. Forthcoming. "The Global Tax Deal and the New International Economic Governance." *Tax Law Review*. https://papers.ssrn.com/sol3/papers.cfm?abstract_id=4831166#.

Lachapelle, Erick, and Matthew Paterson. 2013. "Drivers of National Climate Policy." *Climate Policy* 13 (5): 547–71.

La Hoz Theuer, Stephanie, Lambert Schneider, and Derik Broekhoff. 2019. "When Less Is More: Limits to International Transfers under Article 6 of the Paris Agreement." *Climate Policy* 19 (4): 401–13.

Lakhani, Nina. 2023. "Revealed: Top Carbon Offset Projects May Not Cut Planet-Heating Emissions." *The Guardian*, September 19. https://www.theguardian.com/environment/2023/sep/19/do-carbon-credit-reduce-emissions-greenhouse-gases (accessed October 22, 2024).

Lazarus, Oliver, Sonali McDermid, and Jennifer Jacquet. 2021. "The Climate Responsibilities of Industrial Meat and Dairy Producers." *Climatic Change* 165 (1): 30.

Lee, Carrie M., Michael Lazarus, Gordon R. Smith, Kimberly Todd, and Melissa Weitz. 2013. "A Ton Is Not Always a Ton: A Road-Test of Landfill, Manure, and Afforestation/Reforestation Offset Protocols in the US Carbon Market." *Environmental Science & Policy* 33: 53–62.

Legislative Analyst's Office (LAO). 2010. "Proposition 26: Increases Legislative Vote Requirement to Two-Thirds for State Levies and Charges. Imposes Additional Requirement for
Voters to Approve Local Levies and Charges with Limited Exceptions. Initiative Constitutional Amendment." Sacramento, CA: LAO (July 15). https://lao.ca.gov/ballot/2010/26
_11_2010.aspx (accessed November 4, 2024).

Levin, Kelly, Benjamin Cashore, Steven Bernstein, and Graeme Auld. 2012. "Overcoming the
Tragedy of Super Wicked Problems: Constraining Our Future Selves to Ameliorate Global
Climate Change." *Policy Sciences* 45 (2): 123–52.

Levy, Marc A. 1993. "European Acid Rain: The Power of Tote-Board Diplomacy." In *Institutions
for the Earth: Sources of Effective International Environmental Protection*, edited by Peter M.
Haas, Robert O. Keohane, and Marc A. Levy. Cambridge, MA: MIT Press.

Lewis, Joanna I. 2014. "The Rise of Renewable Energy Protectionism: Emerging Trade Conflicts
and Implications for Low Carbon Development." *Global Environmental Politics* 14 (4):
10–35.

———. 2024. "The Climate Risk of Green Industrial Policy." *Current History* 123 (849): 14–19.

Li, Mei, Gregory Trencher, and Jusen Asuka. 2022. "The Clean Energy Claims of BP, Chevron,
ExxonMobil, and Shell: A Mismatch between Discourse, Actions, and Investments." *PLOS
ONE* 17 (2): e0263596.

Lim, Junghyun, Michaël Aklin, and Morgan R. Frank. 2023. "Location Is a Major Barrier for
Transferring US Fossil Fuel Employment to Green Jobs." *Nature Communications* 14 (1): 5711.

Lockwood, Matthew. 2015. "The Political Dynamics of Green Transformations: Feedback Effects and Institutional Context." In *The Politics of Green Transformations*, edited by Ian
Scoones, Peter Newell, and Melissa Leach. London: Routledge.

Lohmann, Larry. 2008. "Carbon Trading, Climate Justice, and the Production of Ignorance: Ten
Examples." *Development* 51 (3): 359–65.

Lund, Emma. 2010. "Dysfunctional Delegation: Why the Design of the CDM's Supervisory
System Is Fundamentally Flawed." *Climate Policy* 10 (3): 277–88.

Lund, Jens Friis, Nils Markusson, Wim Carton, and Holly Jean Buck. 2023. "Net Zero and the
Unexplored Politics of Residual Emissions." *Energy Research & Social Science* 98: 103035.

Luo, Huilin, Wei Peng, Allen Fawcett, Jessica Green, Gokul Iyer, Jonas Meckling, and David G.
Victor. 2024. "Modeling the Impacts of Policy Sequencing on Energy Decarbonization."
Unpublished manuscript.

Lyons, Kate. 2022. "How to Move a Country: Fiji's Radical Plan to Escape Rising Sea Levels."
The Guardian, November 8. https://www.theguardian.com/environment/2022/nov/08
/how-to-move-a-country-fiji-radical-plan-escape-rising-seas-climate-crisis (accessed October 30, 2024).

Lyons, Kristen, and Peter Westoby. 2014. "Carbon Colonialism and the New Land Grab: Planta-
tion Forestry in Uganda and Its Livelihood Impacts." *Journal of Rural Studies* 36: 13–21.

Major, Darren. 2023. "Ottawa Exempting Home Heating Oil from Carbon Tax for 3 Years,
Trudeau Says." *CBC News*, October 26. https://www.cbc.ca/news/politics/trudeau-pause
-carbon-tax-rural-home-heating-1.7009347 (accessed October 18, 2024).

Malm, Andreas. 2016. *Fossil Capital: The Rise of Steam Power and the Roots of Global Warming*.
London: Verso.

Malo, Sebastien. 2019. "UN Reform Needed to Stop Companies Fighting Climate Rules: Nobel
Laureate Stiglitz." *Reuters*, May 29. https://www.reuters.com/article/world/un-reform
-needed-to-stop-companies-fighting-climate-rules-nobel-laureate-stig-idUSKCN1SZ04X
/ (accessed October 28, 2024).

Marland, Gregg. 2008. "Uncertainties in Accounting for CO_2 from Fossil Fuels." *Journal of Industrial Ecology* 12 (2): 136–39.

Marland, Gregg, Khrystyna Hamal, and Matthias Jonas. 2009. "How Uncertain Are Estimates of CO2 Emissions?" *Journal of Industrial Ecology* 13 (1): 4–7.

Martin, Geoff, and Eri Saikawa. 2017. "Effectiveness of State Climate and Energy Policies in Reducing Power-Sector CO₂ Emissions." *Nature Climate Change* 7 (12): 912–19.

Mastrandrea, Michael D., Mason Inman, and Danny Cullenward. 2020. "Assessing California's Progress toward Its 2020 Greenhouse Gas Emissions Limit." *Energy Policy* 138: 111219.

Mattli, Walter, and Ngaire Woods. 2009. *The Politics of Global Regulation*. Princeton, NJ: Princeton University Press.

Mazzucato, Mariana. 2015. "Innovation, the State, and Patient Capital." *Political Quarterly* 86 (S1): 98–118.

McGivern, Alexis, Kaya Axelsson, Saskia Straub, Sylvie Craig, and Bernard Steen. 2022. "Defining Net Zero for Organisations: How Do Climate Criteria Align across Standards and Voluntary Initiatives?" Oxford: University of Oxford (October). https://netzeroclimate .org/wp-content/uploads/2022/12/Summary-Report_Oxford-Net-Zero_October-2022 .pdf.

McNamara, Kathleen R. 2024. "Transforming Europe? The EU's Industrial Policy and Geopolitical Turn." *Journal of European Public Policy* 31 (9): 2371–96.

Meckling, Jonas. 2011. *Carbon Coalitions: Business, Climate Politics, and the Rise of Emissions Trading*. Cambridge, MA: MIT Press.

———. 2015. "Oppose, Support, or Hedge? Distributional Effects, Regulatory Pressure, and Business Strategy in Environmental Politics." *Global Environmental Politics* 15 (2): 19–37.

———. 2021. "Making Industrial Policy Work for Decarbonization." *Global Environmental Politics* 21 (4): 134–47.

Meckling, Jonas, Nina Kelsey, Eric Biber, and John Zysman. 2015. "Winning Coalitions for Climate Policy." *Science* 349 (6253): 1170–71.

Meckling, Jonas, Phillip Y. Lipscy, Jared J. Finnegan, and Florence Metz. 2022. "Why Nations Lead or Lag in Energy Transitions." *Science* 378 (6615): 31–33.

Mehling, Michael A., Gilbert E. Metcalf, and Robert N. Stavins. 2018a. "Linking Heterogenous Climate Policies (Consistent with the Paris Agreement)." *Environmental Law* 48 (4): 647–98.

———. 2018b. "Linking Climate Policies to Advance Global Mitigation." *Science* 359 (6379): 997–98.

Meckling, Jonas, and Jonas Nahm. 2018. "When Do States Disrupt Industries? Electric Cars and the Politics of Innovation." *Review of International Political Economy* 25 (4): 505–29.

———. 2021. "Strategic State Capacity: How States Counter Opposition to Climate Policy." *Comparative Political Studies* 55 (3): 493–523.

Mehling, Michael A., and Robert A. Ritz. 2020. "Going Beyond Default Intensities in an EU Carbon Border Adjustment Mechanism." Cambridge: University of Cambridge, Energy Policy Research Group (September). https://www.jstor.org/stable/resrep30315 (accessed November 20, 2023).

Mehling, Michael A., Harro van Asselt, Kasturi Das, Susanne Droege, and Cleo Verkuijl. 2019. "Designing Border Carbon Adjustments for Enhanced Climate Action." *American Journal of International Law* 113 (3): 433–81.

Mehling, Michael, Harro van Asselt, Susanne Droege, and Kasturi Das. 2022. "The Form and Substance of International Cooperation on Border Carbon Adjustments." *AJIL Unbound* 116: 213–18.

Mehranvar, Ladan, and Martin Brauch. 2024. "Breaking Free: Strategies for Governments on Terminating Investment Treaties and Removing ISDS Provisions." New York: Columbia Center on Sustainable Investment (October). https://scholarship.law.columbia.edu/cgi /viewcontent.cgi?article=1048&context=sustainable_investment.

Meltzer, Joshua. 2013. "The International Civil Aviation Organization's Regulation of CO2 Emissions: Amending the EU Aviation Directive to Avoid a Trade War." Washington, DC: Brookings Institution (October 25). https://www.brookings.edu/articles/the-international-civil-aviation-organizations-regulation-of-co2-emissions-amending-the-eu-aviation-directive-to-avoid-a-trade-war/ (accessed November 19, 2024).

Meyer, Timothy, and Todd N. Tucker. 2022. "A Pragmatic Approach to Carbon Border Measures." *World Trade Review* 21 (1): 109–20.

Michaelowa, Katharina, and Axel Michaelowa. 2012. "Negotiating Climate Change." *Climate Policy* 12 (5): 527–33.

Mildenberger, Matto. 2020. *Carbon Captured: How Business and Labor Control Climate Politics.* Cambridge, MA: MIT Press.

Mildenberger, Matto, Erick Lachapelle, Kathryn Harrison, and Isabelle Stadelman-Steffen. 2022. "Limited Impacts of Carbon Tax Rebate Programmes on Public Support for Carbon Pricing." *Nature Climate Change* 12 (2): 141–47.

Mildenberger, Matto, and Leah Stokes. 2020. "The Trouble with Carbon Pricing." *Boston Review,* September 24. http://bostonreview.net/science-nature-politics/matto-mildenberger-leah-c-stokes-trouble-carbon-pricing (accessed September 25, 2020).

Mitchell, Timothy. 2013. *Carbon Democracy: Political Power in the Age of Oil.* London: Verso.

Morse, Ian. 2023. "Inside the Little-Known Group Setting the Corporate Climate Agenda." *MIT Technology Review,* May 16. https://www.technologyreview.com/2023/05/16/1073064/inside-the-little-known-group-setting-the-corporate-climate-agenda/ (accessed February 20, 2024).

Murphy, Richard. 2015. *The Joy of Tax: How a Fair Tax System Can Create a Better Society.* London: Bantam Press.

Murray, Brian, and Nicholas Rivers. 2015. "British Columbia's Revenue-Neutral Carbon Tax: A Review of the Latest 'Grand Experiment' in Environmental Policy." *Energy Policy* 86: 674–83.

Murray, James. 2012. "Russia Fires First Shot in EU Aviation Emissions Trade War." *The Guardian,* February 22. https://www.theguardian.com/environment/2012/feb/22/russia-eu-aviation-emissions-trade (accessed November 19, 2024).

Nahm, Jonas. 2021. *Collaborative Advantage: Forging Green Industries in the New Global Economy.* New York: Oxford University Press.

Nakićenović, Nebojša, and Aviott John. 1991. "CO2 Reduction and Removal: Measures for the Next Century." *Energy* 16 (11): 1347–77.

Nemes, Noémi, Stephen J. Scanlan, Pete Smith, Tone Smith, Melissa Aronczyk, Stephanie Hill, et al. 2022. "An Integrated Framework to Assess Greenwashing." *Sustainability* 14 (8): 4431.

Neslen, Arthur. 2023. "EU, Germany, and Denmark Sued by Oil Firm over Windfall Tax." *The Guardian,* November 20. https://www.theguardian.com/world/2023/nov/20/eu-germany-and-denmark-sued-by-oil-firm-over-windfall-tax (accessed October 28, 2024).

Net Zero Regulation and Policy Hub. 2023. "Net Zero Regulation Stocktake: What Does the Road to Net Zero Look Like?" Oxford: University of Oxford. https://netzeroclimate.org/regulation-tracking/.

Neville, Kate J. 2020. "Shadows of Divestment: The Complications of Diverting Fossil Fuel Finance." *Global Environmental Politics* 20 (2): 3–11.

NewClimate Institute, Oxford Net Zero, Energy & Climate Intelligence Unit, and Data-Driven EnviroLab. 2023. *Net Zero Stocktake 2023.* Oxford: Oxford University (June). https://ca1-nzt.edcdn.com/Reports/Net_Zero_Stocktake_2023.pdf?v=1696255114 (accessed February 26, 2024).

Newell, Peter, and Matthew Paterson. 2010. *Climate Capitalism: Global Warming and the Transformation of the Global Economy.* Cambridge: Cambridge University Press.

Newell, Peter, and Andrew Simms. 2020. "Towards a Fossil Fuel Non-Proliferation Treaty." *Climate Policy* 20 (8): 1043–54.

Newman, Rebecca, and Ilan Noy. 2023. "The Global Costs of Extreme Weather That Are Attributable to Climate Change." *Nature Communications* 14 (1): 6103.

New Zealand Foreign Affairs & Trade. 2024. "Joint Ministerial Statement on Conclusion of Negotiations for the Agreement on Climate Change, Trade and Sustainability." Wellington: New Zealand Foreign Affairs & Trade (July 2). https://www.mfat.govt.nz/en/media-and-resources/joint-ministerial-statement-on-conclusion-of-negotiations-for-the-agreement-on-climate-change-trade-and-sustainability

Nordhaus, William. 1994. *Managing the Global Commons: The Economics of Climate Change.* Cambridge: MIT Press.

———. 2015. "Climate Clubs: Overcoming Free-Riding in International Climate Policy." *American Economic Review* 105 (4): 1339–70.

Nottage, Luke. 2024. "Australia's Ambivalence Again around Investor-State Arbitration: Comparisons with Europe and Implications for Asia." *ICSID Review: Foreign Investment Law Journal* 39 (August 7): siae029. https://ssrn.com/abstract=4891441.

Office of the United States Trade Representative. 2020. "Agreement between the United States of America, the United Mexican States, and Canada." July 1. https://ustr.gov/trade-agreements/free-trade-agreements/united-states-mexico-canada-agreement/agreement-between.

O'Hara, Clare, and Jeffrey Jones. 2023. "Climate Change Is Making Insurance More Expensive and More Limited—and It's Only Going to Get Worse." *Globe and Mail*, December 1. https://www.theglobeandmail.com/business/article-insurance-coverage-climate-change/ (accessed April 23, 2024).

Ontario Office of the Premier. 2024. "Honda to Build Canada's First Comprehensive Electric Vehicle Supply Chain, Creating Thousands of New Jobs in Ontario." April 25. https://news.ontario.ca/en/release/1004485/honda-to-build-canadas-first-comprehensive-electric-vehicle-supply-chain-creating-thousands-of-new-jobs-in-ontario (accessed November 13, 2024).

Oreskes, Naomi, and Erik M. Conway. 2011. *Merchants of Doubt: How a Handful of Scientists Obscured the Truth on Issues from Tobacco Smoke to Climate Change.* New York: Bloomsbury Publishing.

Organization for Economic Cooperation and Development (OECD). 2023. "The Pillar Two Rules in a Nutshell." Paris: OECD (July). https://www.oecd.org/content/dam/oecd/en/topics/policy-sub-issues/global-minimum-tax/pillar-two-model-rules-in-a-nutshell.pdf.

———. 2024a. *Climate Finance Provided and Mobilised by Developed Countries in 2013–2022.* Paris: OECD Publishing (May 29).

———. 2024b. *Harnessing Public Procurement for the Green Transition.* Paris: OECD Publishing (June 26). https://www.oecd.org/en/publications/2024/06/harnessing-public-procurement-for-the-green-transition_ef16c8d4.html (accessed October 7, 2024).

———. 2024c. *Infrastructure for a Climate-Resilient Future.* Paris: OECD. https://doi.org/10.1787/a74a45b0-en.

———. n.d. "Investment Treaties." https://www.oecd.org/en/topics/sub-issues/the-future-of-investment-treaties.html.

Ørsted. n.d. "In Numbers: Powering the World with Green Energy." https://orsted.com/en/who-we-are/our-purpose/powering-the-world-with-green-energy.

Ostrom, Elinor. 2010a. "Beyond Markets and States: Polycentric Governance of Complex Economic Systems." *American Economic Review* 100 (3): 641–72.

———. 2010b. "Polycentric Systems or Coping with Collective Action and Global Environmental Change." *Global Environmental Change* 20 (special issue, 4): 550–57.

Overdevest, Christine, and Jonathan Zeitlin. 2014. "Assembling an Experimentalist Regime: Transnational Governance Interactions in the Forest Sector." *Regulation & Governance* 8 (1): 22–48.

Owen, David. 2023. "The Great Electrician Shortage." *The New Yorker*, April 24. https://www
.newyorker.com/news/dept-of-energy/the-great-electrician-shortage (accessed November 22, 2024).

Oxfam. 2023. *Climate Equality: A Planet for the 99%.* Oxford: Oxfam (November 20). https://
policy-practice.oxfam.org/resources/climate-equality-a-planet-for-the-99-621551/ (accessed November 21, 2023).

Oxfam and Institute for European Environmental Policy. 2021. "Carbon Inequality in 2030." Oxford: Oxfam. https://oxfamilibrary.openrepository.com/bitstream/handle/10546
/621305/bn-carbon-inequality-2030-051121-en.pdf.

Oxfam and Stockholm Envior. 2020. "The Carbon Inequality Era: An Assessment of the Global Distribution of Consumption Emissions among Individuals from 1990 to 2015 and Beyond. September 21. https://policy-practice.oxfam.org/resources/the-carbon-inequality-era-an
-assessment-of-the-global-distribution-of-consumpti-621049/ (accessed February 20, 2025).

Paddison, Laura. 2017. "Exxon, Shell, and Other Carbon Producers Sued for Sea Level Rises in California." *The Guardian*, July 26. https://www.theguardian.com/sustainable-business
/2017/jul/26/california-communities-lawsuit-exxon-shell-climate-change-carbon-majors
-sea-level-rises (accessed November 1, 2024).

Padilla, Emilio, and Alfredo Serrano. 2006. "Inequality in CO2 Emissions across Countries and Its Relationship with Income Inequality: A Distributive Approach." *Energy Policy* 34 (14): 1762–72.

Paine, Joshua. 2023. "Submission to International Trade Committee Inquiry on UK Trade Negotiations—Windfall Taxes and Investment Treaties." UKT0048 (January). Presented at the UK Parliament, International Trade Committee.

Parry, Ian W. H., Simon Black, and Karlygash Zhunussova. 2022. "Carbon Taxes or Emissions Trading Systems?: Instrument Choice and Design." Washington, DC: World Bank (July 21). https://www.imf.org/en/Publications/staff-climate-notes/Issues/2022/07/14/Carbon
-Taxes-or-Emissions-Trading-Systems-Instrument-Choice-and-Design-519101 (accessed November 15, 2023).

Paterson, Matthew. 2012. "Who and What Are Carbon Markets For? Politics and the Development of Climate Policy." *Climate Policy* 12 (1): 82–97.

———. 2021a. *In Search of Climate Politics.* New York: Cambridge University Press.

———. 2021b. "Climate Change and International Political Economy: Between Collapse and Transformation." *Review of International Political Economy* 28 (2): 394–405.

Paterson, Matthew, and Johannes Stripple. 2012. "Virtuous Carbon." *Environmental Politics* 21 (4): 563–82.

Pauer, Stefan U. 2018. "Including Electricity Imports in California's Cap-and-Trade Program: A Case Study of a Border Carbon Adjustment in Practice." *Electricity Journal* 31 (10, special issue): 39–45.

Peinhardt, Clint, and Rachel L. Wellhausen. 2016. "Withdrawing from Investment Treaties but Protecting Investment." *Global Policy* 7 (4): 571–76.

Petek, Gabriel. 2020. *Assessing California's Climate Policies—Electricity Generation.* Sacramento: Legislative Analyst's Office (January 6). https://lao.ca.gov/Publications/Report/4131.

———. 2023a. "Assessing California's Climate Policies: The 2022 Scoping Plan Update." Sacramento: Legislative Analyst's Office (January 4). https://lao.ca.gov/Publications/Report
/4656.

Petek, Gabriel. 2023b. "California's Cap-and-Trade Program: Frequently Asked Questions." Sacramento: Legislative Analyst's Office (October). https://lao.ca.gov/reports/2023/4811/Cap-and-Trade-FAQs-102423.pdf.

Peters, Glen P. 2008. "From Production-Based to Consumption-Based National Emission Inventories." *Ecological Economics* 65 (1): 13–23.

Peters, Glen P., and Edgar G. Hertwich. 2008. "Post-Kyoto Greenhouse Gas Inventories: Production versus Consumption." *Climatic Change* 86 (1): 51–66.

Potter, Christopher, Vanessa Brooks Genovese, Steven Klooster, Matthew Bobo, and Alicia Torregrosa. 2001. "Biomass Burning Losses of Carbon Estimated from Ecosystem Modeling and Satellite Data Analysis for the Brazilian Amazon Region. *Atmospheric Environment* 35 (10): 1773–81. https://doi.org/10.1016/S1352-2310(00)00459-3.

Prentice, I. Colin, Martin Heimann, and Stephen Sitch. 2000. "The Carbon Balance of the Terrestrial Biosphere: Ecosystem Models and Atmospheric Observations." *Ecological Applications* 10 (6): 1553–73.

Prete, Chiara Lo, Ashish Tyagi, and Qingyu Xu. 2023. "California's Cap-and-Trade Program and Emission Leakage in the Western Interconnection: Comparing Econometric and Partial Equilibrium Model Estimates." *Journal of the Association of Environmental and Resource Economists* 11 (2): 359–402. https://www.journals.uchicago.edu/doi/abs/10.1086/726053 (accessed November 3, 2023).

Pretis, Felix. 2022. "Does a Carbon Tax Reduce CO_2 Emissions? Evidence from British Columbia." *Environmental and Resource Economics* 83 (1): 115–44.

Probst, Benedict S., Malte Toetzke, Andreas Kontoleon, Laura Díaz Anadón, Jan C. Minx, Barbara K. Haya, Lambert Schneider, et al. 2024. "Systematic Assessment of the Achieved Emission Reductions of Carbon Crediting Projects." *Nature Communications* 15 (1): 9562.

PwC. 2025. "PwC's Pillar Two Country Tracker." January 21. https://www.pwc.com/gx/en/tax/international-tax-planning/pillar-two/pwc-pillar-two-country-tracker-summary-v2.pdf.

Rabe, Barry G. 2018. *Can We Price Carbon?* Cambridge, MA: MIT Press.

Race to Zero Expert Peer Review Group. 2022. "Interpretation Guide, Version 2.0." Bonn: United Nations Framework Convention on Climate Change (June). https://climatechampions.unfccc.int/wp-content/uploads/2022/09/EPRG-interpretation-guide.pdf.

Rafaty, Ryan, Geoffroy Dolphin, and Felix Pretis. 2020. "Carbon Pricing and the Elasticity of CO_2 Emissions." Working paper 140. New York: Institute for New Economic Thinking (October 21). https://doi.org/10.36687/inetwp140.

Ranald, Patricia. 2024. "Clive Palmer's Claims against Australia for Billions Renew Pressure to Remove Investor Rights to Sue Governments from Trade Agreements." *Economic and Labour Relations Review* 35 (2): 454–66.

Randazzo, Nina A., Doria R. Gordon, and Steven P. Hamburg. 2023. "Improved Assessment of Baseline and Additionality for Forest Carbon Crediting." *Ecological Applications* 33 (3): e2817.

Ranjan, Prabhash. 2023. "Investor-State Dispute Settlement and Tax Matters: Limitations on State's Sovereign Right to Tax." *Asia Pacific Law Review* 31 (1): 219–34.

Ranson, Matthew, and Robert N. Stavins. 2016. "Linkage of Greenhouse Gas Emissions Trading Systems: Learning from Experience." *Climate Policy* 16 (3): 284–300.

Raymond, Leigh. 2020. "Carbon Pricing and Economic Populism: The Case of Ontario." *Climate Policy* 20 (9): 1127–40.

Rekker, Saphira, M. C. Ives, B. Wade, L. Webb, and C. Greig. 2022. "Measuring Corporate Paris Compliance Using a Strict Science-Based Approach." *Nature Communications* 13 (1): 4441.

Rennert, Kevin, Frank Errickson, Brian C. Prest, Lisa Rennels, Richard G. Newell, William Pizer, et al. 2022. "Comprehensive Evidence Implies a Higher Social Cost of CO_2." *Nature* 610 (7933): 687–92.

Reuters. 2023. "EU's Dombrovskis to Visit US for Talks on Inflation Reduction Act." *Reuters*, February 23. https://www.reuters.com/world/eus-dombrovskis-visit-us-talks-inflation -reduction-act-2023-02-23/ (accessed November 19, 2024).

———. 2024. "China to Crack Down on Emissions Data Fraud as CO_2 Market Expansion Nears." *Reuters*, February 5. https://www.reuters.com/sustainability/boards-policy -regulation/china-crack-down-emissions-data-fraud-co2-market-expansion-nears-2024-02 -05 (accessed October 18, 2024).

Richstein, Jörn C., Émile J. L. Chappin, and Laurens J. de Vries. 2015. "The Market (In-)stability Reserve for EU Carbon Emission Trading: Why It Might Fail and How to Improve It." *Utilities Policy* 35: 1–18.

Ricke, Katharine, Laurent Drouet, Ken Caldeira, and Massimo Tavoni. 2018. "Country-Level Social Cost of Carbon." *Nature Climate Change* 8 (10): 895–900.

Riofrancos, Thea. 2020. *Resource Radicals: From Petro-Nationalism to Post-Extractivism in Ecuador*. Durham, NC: Duke University Press.

Ritchie, Hannah, Pablo Rosado, and Max Roser. 2023. "Per Capita, National, Historical: How Do Countries Compare on CO2 Metrics?" Our World in Data, September 26. https:// ourworldindata.org/co2-emissions-metrics (accessed April 26. 2024).

Rödenbeck, C., Sander Houweling, Manuel Gloor, and Martin Heimann. 2003. "CO2 Flux History 1982–2001 Inferred from Atmospheric Data Using a Global Inversion of Atmospheric Transport." *Atmospheric Chemistry and Physics* 3(6): 1919–64. https://doi.org/10 .5194/acp-3-1919-2003.

Rodrik, Dani. 2019. "Globalization's Wrong Turn: And How It Hurt America." *Foreign Affairs* 98 (4): 26–33.

———. 2022. *An Industrial Policy for Good Jobs*. Washington, DC: The Hamilton Project (September 28). https://www.hamiltonproject.org/publication/policy-proposal/an-industrial -policy-for-good-jobs/.

Rogelj, Joeri, Piers M. Forster, Elmar Kriegler, Christopher J. Smith, and Roland Séférian. 2019. "Estimating and Tracking the Remaining Carbon Budget for Stringent Climate Targets." *Nature* 571 (7765): 335–42.

Rogelj, Joeri, Oliver Geden, Annette Cowie, and Andy Reisinger. 2021. "Net-Zero Emissions Targets Are Vague: Three Ways to Fix." *Nature* 591 (7850): 365–68.

Romero, Christy Goldsmith. 2023. "Opening Statement of Commissioner Christy Goldsmith Romero: The CFTC's Role with Voluntary Carbon Credit Markets." Washington, DC: US Commodity Futures Trading Commission (July 19). https://www.cftc.gov/PressRoom /SpeechesTestimony/romerostatement071923b (accessed October 27, 2024).

Romm, Joseph. 2023. "Are Carbon Offsets Unscalable, Unjust, and Unfixable—and a Threat to the Paris Climate Agreement?" Philadelphia: Penn Center for Science, Sustainability, and the Media. https://bpb-us-w2.wpmucdn.com/web.sas.upenn.edu/dist/0/896/files/2023 /06/OffsetPaper7.0-6-27-23-FINAL2.pdf (accessed September 27, 2023).

Rosenbloom, Daniel, Jochen Markard, Frank W. Geels, and Lea Fuenfschilling. 2020. "Why Carbon Pricing Is Not Sufficient to Mitigate Climate Change—and How 'Sustainability Transition Policy' Can Help." *Proceedings of the National Academy of Sciences* 117 (16): 8664– 68. https://www.pnas.org/content/early/2020/04/07/2004093117 (accessed April 15, 2020).

Rumble, Olivia. 2024. "Zimbabwe Seeks to Calm Carbon Market Waters." *African Climate Wire*, August 15. https://africanclimatewire.org/2024/08/zimbabwe-seeks-to-calm-carbon -market-waters.

Sabin Center for Climate Change Law. 2022. "Asmania et al. vs. Holcim." New York: Sabin Center for Climate Change Law. https://climatecasechart.com/non-us-case/four-islanders -of-pari-v-holcim (accessed November 3, 2024).

Saez, Emmanuel, and Gabriel Zucman. 2019. *The Triumph of Injustice: How the Rich Dodge Taxes and How to Make Them Pay.* New York: W. W. Norton.

Salway, Hugh. 2021. "Comment: We're Still In—Let's Align the Voluntary Carbon Market with Paris Rather than Play by Our Own Rules." Châtelaine, Switzerland: Gold Standard (February 19). https://www.goldstandard.org/news/comment-were-still-in--lets-align-the-voluntary.

Sandler, Todd. 2004. *Global Collective Action.* Cambridge: Cambridge University Press.

Schelling, Thomas C. 1997. "The Cost of Combating Global Warming." *Foreign Affairs* (November/December). https://www.foreignaffairs.com/articles/1997-11-01/cost-combating-global-warming (accessed January 20, 2021).

Schneider, Lambert. 2009. "Assessing the Additionality of CDM Projects: Practical Experiences and Lessons Learned." *Climate Policy* 9 (3): 242–54.

———. 2011. "Perverse Incentives under the CDM: An Evaluation of HFC-23 Destruction Projects." *Climate Policy* 11 (2): 851–64.

Schneider, Lambert, Anja Kollmuss, and Michael Lazarus. 2015. "Addressing the Risk of Double Counting Emission Reductions under the UNFCCC." *Climatic Change* 131 (4): 473–86.

Schneider, Laura C., Ann P. Kinzig, Eric D. Larson, and Luis A. Solórzano. 2001. "Method for Spatially Explicit Calculations of Potential Biomass Yields and Assessment of Land Availability for Biomass Energy Production in Northeastern Brazil." *Agriculture, Ecosystems & Environment* 84 (3): 207–26.

Schwarze, Reimund. 2000. "Activities Implemented Jointly: Another Look at the Facts." *Ecological Economics* 32 (2): 255–67.

Science Based Targets Initiative. 2023. "Corporate Climate Action Gets a Boost with Upgrade to Target Validation and Standard Setting." London: SBTi (September 13). https://sciencebasedtargets.org/news/corporate-climate-action-gets-a-boost-with-upgrade-to-target-validation-and-standard-setting (accessed November 4, 2024).

———. 2024a. "SBTi Doubles Corporate Climate Validations in One Year as Scale Up Gathers Pace." London: SBTi (January 30). https://sciencebasedtargets.org/news/sbti-scale-up-gathers-pace (accessed November 4, 2024).

———. 2024b. "Corporate Net-Zero Standard Criteria, Version 1.2." London: SBTi (March). https://sciencebasedtargets.org/resources/files/Net-Zero-Standard-Criteria.pdf.

———. 2004c. "SBTi Releases Plans for the Corporate Net-Zero Standard Major Revision." London: SBTi (May 9). https://sciencebasedtargets.org/news/sbti-releases-plans-for-the-corporate-net-zero-standard-major-revision.

Semieniuk, Gregor, Philip B. Holden, Jean-Francois Mercure, Pablo Salas, Hector Pollitt, Katharine Jobson, et al. 2022. "Stranded Fossil-Fuel Assets Translate to Major Losses for Investors in Advanced Economies." *Nature Climate Change* 12 (6): 532–38.

Sherrington, Rachel, Clare Carlile, and Hazel Healy. 2023. "Big Meat and Dairy Lobbyists Turn Out in Record Numbers at Cop28." *The Guardian*, December 8. https://www.theguardian.com/environment/2023/dec/09/big-meat-dairy-lobbyists-turn-out-record-numbers-cop28 (accessed November 1, 2024).

Si, Yutong, Dipa Desai, Diana Bozhilova, Sheila Puffer, and Jennie C. Stephens. 2023. "Fossil Fuel Companies' Climate Communication Strategies: Industry Messaging on Renewables and Natural Gas." *Energy Research & Social Science* 98: 103028.

Sinha, Apra, Ashish Kumar Sedai, Abhishek Kumar, and Rabindra Nepal. 2023. "Are Autocracies Bad for the Environment? Global Evidence from Two Centuries of Data." *Energy Journal* 44 (2): 47–78.

Skjaerseth, Jon Birger, and Tora Skodvin. 2003. *Climate Change and the Oil Industry: Common Problems, Different Strategies.* Manchester: Manchester University Press.

Skjaerseth, Jon Birger, and Jørgen Wettestad. 2016. *EU Emissions Trading: Initiation, Decision-making and Implementation*. London: Routledge.

Slaughter, Anne Marie. 2004. *A New World Order*. Princeton, NJ: Princeton University Press.

Sorbe, Stéphane, Peter Gal, and Valentine Millot. 2018. "Can Productivity Still Grow in Service-Based Economies? Literature Overview and Preliminary Evidence from OECD Countries." Paris: OECD (December 21). https://www.oecd-ilibrary.org/economics/can-productivity-still-grow-in-service-based-economies_4458ec7b-en (accessed May 3, 2024).

Stapp, Jared, Christoph Nolte, Matthew Potts, Matthias Baumann, Barbara K. Haya, and Van Butsic. 2023. "Little Evidence of Management Change in California's Forest Offset Program." *Communications Earth & Environment* 4 (1): 1–10.

Stavins, Robert N. 2011. "The Problem of the Commons: Still Unsettled after 100 Years." *American Economic Review* 101 (1): 81–108.

Stern, Nicholas. 2007. *The Economics of Climate Change: The Stern Review*. Cambridge: Cambridge University Press.

Stewart, Richard B., Michael Oppenheimer, and Bryce Rudyk. 2013. "A New Strategy for Global Climate Protection." *Climatic Change* 120 (1): 1–12.

Stiell, Simon. 2024. "New UN Climate Change Report Shows National Climate Plans 'Fall Miles Short of What's Needed.'" United Nations: Climate Change, October 28. https://unfccc.int/news/new-un-climate-change-report-shows-national-climate-plans-fall-miles-short-of-what-s-needed.

Stiglitz, Joseph, and Nicholas Stern. 2017. "Report of the High-Level Commission on Carbon Prices." Washington DC: World Bank (May 29). https://www.carbonpricingleadership.org/report-of-the-highlevel-commission-on-carbon-prices (accessed February 20, 2025).

Stoddard, Isak, Kevin Anderson, Stuart Capstick, Wim Carton, Joanna Depledge, Keri Facer, et al. 2021. "Three Decades of Climate Mitigation: Why Haven't We Bent the Global Emissions Curve?" *Annual Review of Environment and Resources* 46 (1): 653–89.

Stokes, Leah Cardamore. 2020. *Short Circuiting Policy: Interest Groups and the Battle over Clean Energy and Climate Policy in the American States*. New York: Oxford University Press.

Stokes, Leah C., and Hanna L. Breetz. 2018. "Politics in the US Energy Transition: Case Studies of Solar, Wind, Biofuels, and Electric Vehicles Policy." *Energy Policy* 113: 76–86.

Stokes, Leah, and Matto Mildenberger. 2020. "The Trouble with Carbon Pricing." *Boston Review*, September 24. http://bostonreview.net/science-nature-politics/matto-mildenberger-leah-c-stokes-trouble-carbon-pricing (accessed September 25, 2020).

Stroup, Sarah S., and Wendy H. Wong. 2018. "Authority, Strategy, and Influence: Environmental INGOs in Comparative Perspective." *Environmental Politics* 27 (6): 1101–21.

Sultana, Farhana. 2022. "Critical Climate Justice." *Geographical Journal* 188 (1): 118–24.

Supran, Geoffrey, and Naomi Oreskes. 2017. "Assessing ExxonMobil's Climate Change Communications (1977–2014)." *Environmental Research Letters* 12 (8): 084019.

———. 2021. "Rhetoric and Frame Analysis of ExxonMobil's Climate Change Communications." *One Earth* 4 (5): 696–719.

Sutton, Trevor, and Mike Williams. 2023. "Trade beyond Neoliberalism: Concluding a Global Arrangement on Sustainable Steel and Aluminum." Washington, DC: Center for American Progress (December 4). https://www.americanprogress.org/article/trade-beyond-neoliberalism-concluding-a-global-arrangement-on-sustainable-steel-and-aluminum/ (accessed October 8, 2024).

Swyngedouw, Erik. 2010. "Apocalypse Forever?" *Theory, Culture & Society* 27 (2/3): 213–32.

Syed, Fatima. 2021. "Canada's Supreme Court Rules Carbon Price Constitutional. Here's What You Need to Know." *The Narwhal*, March 25. https://thenarwhal.ca/carbon-tax-supreme-court-canada/.

Táíwò, Olúfémi O. 2022. *Reconsidering Reparations*. New York: Oxford University Press.

Tasker, John Paul. 2021. "Trudeau Calls for Global Carbon Tax at COP26 Summit." *CBC News*, November 2. https://www.cbc.ca/news/politics/trudeau-carbon-tax-global-1.6233936 (accessed October 30, 2024).

Temple, James, and Lisa Song. 2021. "The Climate Solution Adding Millions of Tons of CO_2 into the Atmosphere." *MIT Technology Review*, April 29. https://www.technologyreview.com/2021/04/29/1017811/california-climate-policy-carbon-credits-cause-co2-pollution/ (accessed December 5, 2023).

Thompson, Alexander. 2006. "Management under Anarchy: The International Politics of Climate Change." *Climatic Change* 78 (1): 7–29.

———. 2010. "Rational Design in Motion: Uncertainty and Flexibility in the Global Climate Regime." *European Journal of International Relations* 16 (2): 269–96.

Thompson, Alexander, Tomer Broude, and Yoram Z. Haftel. 2019. "Once Bitten, Twice Shy? Investment Disputes, State Sovereignty, and Change in Treaty Design." *International Organization* 73 (4): 859–80.

Tienhaara, Kyla. 2009. *The Expropriation of Environmental Governance: Protecting Foreign Investors at the Expense of Public Policy.* Cambridge: Cambridge University Press.

———. 2018. "Regulatory Chill in a Warming World: The Threat to Climate Policy Posed by Investor-State Dispute Settlement." *Transnational Environmental Law* 7 (2): 229–50.

Tienhaara, Kyla, and Lorenzo Cotula. 2020. "Raising the Cost of Climate Action? Investor-State Dispute Settlement and Compensation for Stranded Fossil Fuel Assets." London: International Institute for Environment and Development (IIED). https://www.iied.org/17660iied (accessed November 28, 2023).

Tienhaara, Kyla, Rachel Thrasher, B. Alexander Simmons, and Kevin P. Gallagher. 2022. "Investor-State Disputes Threaten the Global Green Energy Transition." *Science* 376 (6594): 701–3.

———. 2023. "Investor-State Dispute Settlement: Obstructing a Just Energy Transition." *Climate Policy* 23 (9): 1197–1212.

Tingley, Dustin, and Michael Tomz. 2014. "Conditional Cooperation and Climate Change." *Comparative Political Studies* 47 (3): 344–68.

Tol, Richard S. J. 2019. "A Social Cost of Carbon for (Almost) Every Country." *Energy Economics* 83: 555–66.

Tooze, Adam. 2022. "Welcome to the World of the Polycrisis." *Financial Times*, October 28. https://www.ft.com/content/498398e7-11b1-494b-9cd3-6d669dc3de33.

Tørsløv, Thomas, Ludvig Wier, and Gabriel Zucman. 2023. "The Missing Profits of Nations." *Review of Economic Studies* 90 (3): 1499–1534.

Travers, Eileen. 2024. "Why the World Needs a UN Global Tax Convention." *UN News*, August 16. https://news.un.org/en/story/2024/08/1153301 (accessed October 28, 2024).

Trencher, Gregory, Sascha Nick, Jordan Carlson, and Matthew Johnson. 2024. "Demand for Low-Quality Offsets by Major Companies Undermines Climate Integrity of the Voluntary Carbon Market." *Nature Communications* 15 (1): 6863.

Trexler, Mark, and Auden Schendler. 2015. "Science-Based Carbon Targets for the Corporate World: The Ultimate Sustainability Commitment, or a Costly Distraction?" *Journal of Industrial Ecology* 19 (6): 931–33.

Tucker, Todd. 2019a. "Industrial Policy and Planning: What It Is and How to Do It Better." New York: Roosevelt Institute (July 30). https://rooseveltinstitute.org/publications/industrial-policy-and-planning.

———. 2019b. "The Green New Deal: A Ten-Year Window to Reshape International Economic Law?" SSRN, July 30. https://www.ssrn.com/abstract=3411142 (accessed September 9, 2021).

Twidale, Susanna. 2023. "Carbon Offset Firm South Pole Cuts Ties with Zimbabwe Forest Project." *Reuters*, October 27. https://www.reuters.com/sustainability/cop/carbon-offset-firm-south-pole-cuts-ties-with-zimbabwe-forest-project-2023-10-27/ (accessed October 22, 2024).

UK Government. 2019. *The Climate Change Act 2008 (2050 Target Amendment)*. Vol. 2019, no. 1056. Richmond: National Archives, legislation.gov.uk. https://www.legislation.gov.uk/uksi/2019/1056/contents/made.

Unger, Charlotte, and Rainer Quitzow. 2024. "Dream or Reality: Where Is the Club for Green Steel?" *npj Climate Action* 3 (1): 1–3.

UN Environment Programme. 2023. "Emissions Gap Report 2023: Broken Record." November 20. https://www.unep.org/resources/emissions-gap-report-2023 (accessed October 28, 2024).

United Nations Framework Convention on Climate Change (UNFCCC). 2001. "Modalities and Procedures for a Clean Development Mechanism, as Defined in Article 12 of the Kyoto Protocol." Bonn: UNFCCC. https://unfccc.int/files/meetings/workshops/other_meetings/application/pdf/17cp7.pdf.

———. 2009. "Copenhagen Accord." Bonn: UNFCCC (December 18). https://unfccc.int/resource/docs/2009/cop15/eng/l07.pdf.

———. 2015. "Paris Agreement." Bonn: UNFCCC (December 12). https://unfccc.int/process-and-meetings/the-paris-agreement.

———. 2016. "Report of the Conference of the Parties on Its 21st Session." Bonn: UNFCCC (January 29). https://unfccc.int/resource/docs/2015/cop21/eng/10a01.

———. 2018. "Achievements of the Clean Development Mechanism: Harnessing Incentive for Climate Action, 2001–2018." Bonn: UNFCCC. https://unfccc.int/sites/default/files/resource/UNFCCC_CDM_report_2018.pdf.

———. 2021. "Guidance on Cooperative Approaches Referred to in Article 6, Paragraph 2, of the Paris Agreement." Bonn: UNFCCC. https://unfccc.int/sites/default/files/resource/cma2021_10_add1_adv.pdf#page=11 (accessed October 21, 2024).

———. 2022a. "Race to Zero Criteria 3.0." Bonn: UNFCCC. https://racetozero.unfccc.int/system/criteria/.

———. 2022b. "Rules, Modalities, and Procedures for the Mechanism Established by Article 6, Paragraph 4, of the Paris Agreement." Bonn: UNFCCC (March 8). https://unfccc.int/sites/default/files/resource/cma2021_10_add1_adv.pdf#page=25 (accessed September 3, 2024).

———. 2024a. "Report of the Conference of the Parties Serving as the Meeting of the Parties to the Paris Agreement on Its Fifth Session, Held in the United Arab Emirates from 30 November to 13 December 2023." Bonn: UNFCCC (March 15). https://unfccc.int/documents/637072.

———. 2024b. "Rules, Modalities, and Procedures for the Mechanism Established by Article 6, Paragraph 4, of the Paris Agreement and Referred to in Decision 3/CMA.3." Bonn: UNFCCC (November 23). https://unfccc.int/sites/default/files/resource/cma2024_L16E.pdf.

———. 2024c. "Matters Relating to Finance." Bonn: UNFCCC (November 24). https://unfccc.int/sites/default/files/resource/cma2024_L22_adv.pdf.

———. 2024d. "Conference of the Parties Serving as the Meeting of the Parties to the Paris Agreement: Nationally Determined Contributions under the Paris Agreement." Bonn: UNFCCC (October 24). https://unfccc.int/sites/default/files/resource/cma2024_10_adv.pdf?download.

———. 2024e. "UN Climate Change Conference Baku—November 2024: Statistics on Participation and In-Session Engagement." https://unfccc.int/process-and-meetings/parties-non-party-stakeholders/non-party-stakeholders/statistics-on-non-party-stakeholders/statistics-on-participation-and-in-session-engagement.

United Nations Framework Convention on Climate Change (UNFCCC). n.d. "Statistics on Participation and In-Session Engagement." Bonn: UNFCCC. https://unfccc.int/process-and-meetings/parties-non-party-stakeholders/non-party-stakeholders/statistics-on-non-party-stakeholders/statistics-on-participation-and-in-session-engagement (accessed April 23, 2024).

UN General Assembly. 2024a. "Promotion of Inclusive and Effective International Tax Cooperation at the United Nations." United Nations, Ad Hoc Committee to Draft Terms of Reference for a United Nations Framework Convention on International Tax Cooperation, August 30. https://digitallibrary.un.org/record/4062430?ln=en&v=pdf.

———. 2024b. "Chair's Proposal for Draft Terms of Reference for a United Nations Framework Convention on International Tax Cooperation." New York: Ad Hoc Committee to Draft Terms of Reference for a United Nations Framework Convention on International Tax Cooperation (August 15). https://financing.desa.un.org/sites/default/files/2024-08/Chair%27s%20proposal%20draft%20ToR_L.4_15%20Aug%202024____.pdf.

Unruh, Gregory C. 2000. "Understanding Carbon Lock-In." *Energy Policy* 28 (12): 817–30.

US Department of Commerce (DOC). 2024. "Two Years Later: Funding from CHIPS and Science Act Creating Quality Jobs, Growing Local Economies, and Bringing Semiconductor Manufacturing Back to America." Washington, DC: DOC (August 9). https://www.commerce.gov/news/blog/2024/08/two-years-later-funding-chips-and-science-act-creating-quality-jobs-growing-local (accessed November 3, 2024).

US Securities and Exchange Commission (SEC). 2024. "The Enhancement and Standardization of Climate-Related Disclosures for Investors." Updated December 11, 2024. https://www.sec.gov/rules/2022/03/enhancement-and-standardization-climate-related-disclosures-investors (accessed April 29, 2024).

Veenendaal, Elmar M., Olaf Kolle, and Jon Lloyd. 2004. "Seasonal Variation in Energy Fluxes and Carbon Dioxide Exchange for a Broad-Leaved Semi-Arid Savanna (Mopane Woodland) in Southern Africa." *Global Change Biology* 10 (3): 318–28. https://doi.org/10.1111/j.1365-2486.2003.00699.x.

Verbeek, Bart-Jaap. 2023. "The Modernization of the Energy Charter Treaty: Fulfilled or Broken Promises?" *Business and Human Rights Journal* 8 (1): 97–102.

Verra. n.d. "Area of Focus: Blue Carbon." https://verra.org/programs/verified-carbon-standard/area-of-focus-blue-carbon/ (accessed October 27, 2024).

Victor, David. 2001. *The Collapse of the Kyoto Protocol and the Struggle to Slow Global Warming.* Princeton, NJ: Princeton University Press.

———. 2011. *Global Warming Gridlock: Creating More Effective Strategies for Protecting the Planet.* Cambridge: Cambridge University Press.

Victor, David G., Frank W. Geels, and Simon Sharpe. 2019. "Accelerating the Low Carbon Transition: The Case for Stronger, More Targeted and Coordinated International Action." London: Energy Transitions Commission (November). https://www.energy-transitions.org/publications/accelerating-the-low-carbon-transition/ (accessed September 4, 2023).

Victor, David, and Joshua House. 2004. "A New Currency: Climate Change and Carbon Credits." *Harvard International Review* (Summer): 56–59.

Victor, David G., Joshua C. House, and Sarah Joy. 2005. "A Madisonian Approach to Climate Policy." *Science* 309 (5742): 1820–21.

Vogel, David, and Robert A. Kagan, eds. 2004. *The Dynamics of Regulatory Change: How Globalization Affects National Regulatory Policies.* Berkeley: University of California Press.

Vormedal, Irja, and Jonas Meckling. 2024. "How Foes Become Allies: The Shifting Role of Business in Climate Politics." *Policy Sciences* 57 (1): 101–24.

Wainwright, Joel, and Geoff Mann. 2018. *Climate Leviathan: A Political Theory of Our Planetary Future.* London: Verso.

Waldheim, L., and E. Carpentieri. 2000. "Update on the Progress of the Brazilian Wood BIG-GT Demonstration Project." *Journal of Engineering for Gas Turbines and Power* 123 (3): 525–36.

Walmart. n.d. "Suppliers." https://corporate.walmart.com/suppliers (accessed November 4, 2024).

Wang, Alex, Daniel Carpenter-Gold, and Andria So. 2022. "Key Governance Issues in California's Carbon Cap and Trade." Los Angeles: UCLA School of Law, California-China Climate Institute (May). https://ccci.berkeley.edu/sites/default/files/Key_Governance_Issues_in_California_Carbon-Cap-and-Trade_System-Final.pdf.

Wang, Zhongying, Haiyan Qin, and Joanna I. Lewis. 2012. "China's Wind Power Industry: Policy Support, Technological Achievements, and Emerging Challenges." *Energy Policy* 51: 80–88.

Wara, Michael. 2006. "Measuring the Clean Development Mechanism's Performance and Potential." Working Paper 56. Stanford, CA: Stanford University, Program on Energy and Sustainable Development (July 1). https://pesd.fsi.stanford.edu/publications/cdm.

———. 2007. "Is the Global Carbon Market Working?" *Nature* 445 (7128): 595–96.

———. 2014. "California's Energy and Climate Policy: A Full Plate, but Perhaps Not a Model Policy." *Bulletin of the Atomic Scientists* 70 (5): 26–34.

Weikmans, Romain, and J. Timmons Roberts. 2019. "The International Climate Finance Accounting Muddle: Is There Hope on the Horizon?" *Climate and Development* 11 (2): 97–111.

Wenzel, Fernanda. 2024. "Top Brands Buy Amazon Carbon Credits from Suspected Timber Laundering Scam." *Mongabay Environmental News*, May 21. https://news.mongabay.com/2024/05/top-brands-buy-amazon-carbon-credits-from-suspected-timber-laundering-scam/ (accessed October 22, 2024).

Werksman, Jacob. 1998. "The Clean Development Mechanism: Unwrapping the Kyoto Surprise." *Review of European Community and International Environmental Law* 7 (2): 147–58.

Wettestad, Jørgen, and Torbjørg Jevnaker. 2016. *Rescuing EU Emissions Trading: The Climate Policy Flagship*. London: Palgrave Macmillan UK.

The White House. 2024. "Fact Sheet: President Biden Takes Action to Protect American Workers and Businesses from China's Unfair Trade Practices." May 14. https://bidenwhitehouse.archives.gov/briefing-room/statements-releases/2024/05/14/fact-sheet-president-biden-takes-action-to-protect-american-workers-and-businesses-from-chinas-unfair-trade-practices.

Wiedmann, Thomas, Manfred Lenzen, Lorenz T. Keyßer, and Julia K. Steinberger. 2020. "Scientists' Warning on Affluence." *Nature Communications* 11 (1): 3107.

Woodside, John. 2024. "Unpacking Big Oil's Fierce Pushback against New Truth-in-Advertising Rules." *Canada's National Observer*, June 26. https://www.nationalobserver.com/2024/06/26/analysis/unpacking-big-oils-fierce-pushback-against-new-truth-advertising-rules (accessed November 1, 2024).

World Bank. 2022a. *Carbon Pricing Leadership Report, 2021/2022*. Washington, DC: World Bank. https://www.carbonpricingleadership.org/leadershipreports (accessed November 15, 2023).

———. 2022b. "State and Trends of Carbon Pricing 2022." Washington, DC: World Bank. http://hdl.handle.net/10986/37455 (accessed March 11, 2023).

———. 2023a. "State and Trends of Carbon Pricing 2023." Washington, DC: World Bank. http://hdl.handle.net/10986/37455 (accessed March 11, 2023).

———. 2023b. "Global Public Procurement Database: Share, Compare, Improve!" Washington, DC: World Bank (March 23). https://www.worldbank.org/en/news/feature/2020/03/23/global-public-procurement-database-share-compare-improve (accessed November 13, 2024).

World Bank. 2024. "State and Trends of Carbon Pricing 2024." Washington, DC: World Bank. https://openknowledge.worldbank.org/entities/publication/b0d66765-299c-4fb8-921f -61f6bb979087 (accessed September 15, 2024).

———. n.d. "State and Trends of Carbon Pricing Dashboard" (as of November 25, 2024). https://carbonpricingdashboard.worldbank.org.

World Business Council for Sustainable Development and World Resources Institute. 2004. "GHG Protocol Corporate Accounting and Reporting Standard." Washington, DC: World Resources Institute. https://ghgprotocol.org/corporate-standard.

World Meteorological Organization (WMO). 1979. "World Climate Conference: A Conference of Experts on Climate and Mankind." Geneva: WMO (February). https://library.wmo.int /idurl/4/54699.

Yan, Jingchi. 2021. "The Impact of Climate Policy on Fossil Fuel Consumption: Evidence from the Regional Greenhouse Gas Initiative (RGGI)." *Energy Economics* 100: 105333.

Yayasan Auriga Nusantara, Environmental Paper Network, Greenpeace International, Woods and Wayside International, and Rainforest Action Network. 2024. "Deforestation Anonymous: Rainforest Destruction and Social Conflict Driven by PT Mayawana Persada in Indonesian Borneo." Greenpeace International, March. https://issuu.com/greenpeaceinter national/docs/deforestation_anonymous.

Yoder, Kate. 2019. "Oregon Republicans Go into Hiding to Avoid Voting on Climate Bill." *Grist*, June 20. https://grist.org/article/oregon-republicans-go-into-hiding-to-avoid-voting-on -climate-bill (accessed April 25, 2024).

———. 2021. "After a Decade of Failures, Washington State Passes a Cap on Carbon Emissions." *Salon*, May 1. https://www.salon.com/2021/05/01/after-a-decade-of-failures-washington -state-passes-a-cap-on-carbon-emissions_partner/ (accessed April 25, 2024).

Yona, Leehi, Benjamin Cashore, and Mark A. Bradford. 2022. "Factors Influencing the Development and Implementation of National Greenhouse Gas Inventory Methodologies." *Policy Design and Practice* 5 (2): 197–225.

Zhang, Sufang, Philip Andrews-Speed, and Meiyun Ji. 2014. "The Erratic Path of the Low-Carbon Transition in China: Evolution of Solar PV Policy." *Energy Policy* 67: 903–12.

Zhou, Yishu, and Ling Huang. 2021. "How Regional Policies Reduce Carbon Emissions in Electricity Markets: Fuel Switching or Emission Leakage." *Energy Economics* 97: 105209.

Zucman, Gabriel. 2015. *The Hidden Wealth of Nations: The Scourge of Tax Havens*. Chicago: University of Chicago Press.

INDEX

AB32 (Global Warming Solutions Act), 54–55

ACR (American Carbon Registry), 66

activism, 27–28, 95, 109, 133

adaptation, 7, 11, 30, 35, 69, 141–42

additionality: baselines and, 56, 68; of carbon offsets, 68–76; carbon pricing and, 46, 56, 59; greenhouse gases (GHGs) and, 46, 56, 59, 68–76, 79; net zero and, 79

Agreement on Subsidies and Countervailing Measures (WTO), 131

agriculture: asset revaluation and, 32, 34–35; carbon offsets and, 65; carbon pricing and, 44; as fossil asset owners, 32; as green asset owners, 11, 119; greenhouse gas (GHG) emissions and, 10; lobbying and, 10, 34–35; methane and, 10; net zero and, 81; obstructionism of, 10; as vulnerable asset owners, 11

AirCarbon Exchange, 70

Alliance of Small Island States (AOSIS), 35

allowances: asset revaluation and, 27; backloading in emissions trading schemes, 37, 53–54, 117, 121; cap-and-trade schemes and, 43–44, 50, 57, 60; carbon pricing and, 42–44, 50–62; free issuances of, 27, 51, 61, 121, 139; leakage and, 50, 52–53, 55–57, 62; surplus in emission trading schemes, 62

aluminum, 7, 32, 43, 60, 119, 124

Amazon, 86

Amazonian rainforest, 104, 110

Anglo American, 103

Apple, 86

ArcelorMittal, 78

Article 6: Clean Development Mechanism (CDM) and, 59, 65, 67, 75, 138; CORSIA and, 94; double counting and, 69; Gold Standard and, 74; managing tons and, 67, 137–38; Paris Agreement Crediting Mechanism and, 65; Supervisory Body and, 72; two international carbon markets of, 58–59, 62

ASEAN–Australia–New Zealand Free Trade Agreement, 109

Asian Development Bank, 124

asset conversion: asset revaluation and, 22, 31–32, 38; automakers and, 10, 118–19, 126, 129; boundary problem and, 31–33; carbon pricing and, 56; fossil asset owners and, 9, 22, 31–32, 38, 118, 129; green asset owners and, 116, 118, 129; obstructionism and, 9

asset owners: boundary problem and, 31–33; examples of, 29–30; power asymmetry between, 13, 27–8, 35–6, 112–14

asset revaluation: agriculture and, 32, 34; carbon capture and storage (CCS) and, 32; climate change and, 4, 8, 21–30, 34–37; coal industry and, 25, 28, 33–34; convertible industries and, 22, 31–32, 38; decarbonization as driver of, 21–36; as distinct from collective action, 21–24, 27, 29; divestment and, 22, 31, 33; as driver of existential politics, 8–9, 21–38; electricity sector and, 29, 32; fossil asset owners and, 29–34; gas and, 25, 28, 33–34; global climate governance and, 13–17, 28, 37; green asset owners and, 29–33; greenhouse gas (GHG) emissions and, 13, 22, 24–25, 28; greenwashing and, 22, 26, 31, 34; hedging and, 22, 26–27, 31, 36; inequality and, 27, 37; insurance industry and, 29–30; interest groups and, 36–37; international cooperation and, 22–24;

GPSR Authorized Representative: Easy Access System Europe - Mustamäe tee
50, 10621 Tallinn, Estonia, gpsr.requests@easproject.com